Understanding Broadcast and Cable Finance

Understanding Broadcast and Cable Finance

A Primer for the Nonfinancial Manager

Broadcast Cable Financial Management Association (BCFM)

Edited by Walter McDowell, Ph.D., and Alan Batten

Routledge
Taylor & Francis Group
LONDON AND NEW YORK

First published 2008 by Focal Press

Published 2016 by Routledge
4 Park Square, Milton Park, Abingdon, Oxon OX14 4RN
605 Third Avenue, New York, NY 10017

First issued in hardback 2016

Routledge is an imprint of the Taylor and Francis Group, an informa business

Library of Congress Cataloging-in-Publication Data

Understanding broadcast and cable finance : a primer for the nonfinancial manager / Broadcast Cable Financial Management Association (BCFM) ; edited by: Walter McDowell and Alan Batten.
p. cm.
Includes bibliographical references and index.
ISBN-13: 978-0-240-80958-8 (pbk. : alk. paper) 1. Broadcasting–United States–Finance. 2. Cable television–United States–Finance. 3. Broadcasting–United States–Management. 4. Cable television–United States–Management. I. McDowell, Walter. II. Batten, Alan. III. Broadcast Cable Financial Management Association.
HE8689.8.U53 2008
384.54'30681–dc22

2007044889

British Library Cataloguing-in-Publication Data
A catalogue record for this book is available from the British Library.

ISBN 13: 978-1-138-17475-7 (hbk)
ISBN 13: 978-0-240-80958-8 (pbk)

Contents

Foreword

On a hot August afternoon in 1922, Mr. H. M. Blackwell, representing the Queensboro Real Estate Corporation, stood before a microphone within the studios of radio station WEAF in New York City, and read a ten-minute script inviting audiences to visit his new suburban housing development in nearby New Jersey. Within moments after the announcement, the developer's office began receiving dozens of phone calls from people who had listened to the broadcast. Most historians credit this episode as the first known broadcast commercial.

Simultaneously, young men such as William S. Paley and David Sarnoff recognized the commercial viability in this newfangled wireless source of entertainment. Sarnoff had become famous in 1912 for staying at his Marconi wireless set for more than 24 hours to take down the names of the RMS *Titanic*'s survivors. And Paley convinced his cigar-manufacturing father that if radio could sell real estate, it would certainly sell cigars. From those inauspicious starts, today's market-driven amalgam of broadcasting and cable was born.

The obvious success of these ventures and those of other daring businesspeople willing to experiment with this new medium led to the basic advertising business model that would endure for decades to come. Unlike most of the rest of the world, which advocated serious government involvement in the ownership and operation of stations, the founders of U.S. broadcasting preferred an enterprise that was not a government agency, but a private *business*. And, as the saying goes, the rest is history.

The business of radio and television involves not only collecting revenue from advertisers, but also investing revenue in the internal operations of the station, system, or network. Just as car manufacturers must spend money on materials, labor, and dealerships in order to produce and sell a product, so broadcasters must spend money on programming, marketing, facilities, licensing, and personnel in order to attract audiences to sell to advertisers.

The cable industry uses another business model—subscriptions. In addition to selling commercial time to advertisers, cable operators continue to depend on customer subscription fees as their major source of revenue. From basic and premium program networks to video on demand and Internet access, the business of cable is similar to the business of broadcasting in that it involves the proper monitoring and control of money coming in and money going out. A more formal name for these crucial activities is *financial management.*

As complex as the conventional broadcasting and cable business models are today, the future promises even more challenges. With the transition from analog to digital technology, the partitions separating one medium from another are disappearing. For example, most television and radio stations now operate Internet web sites that serve as additional profit centers. In addition, both broadcast and cable networks are exploring video on demand via the Internet as a new revenue stream. Will these new media business ventures use an advertising-based or subscription-based business model, or a combination of both? This blurring of media boundaries, often referred to as media convergence, and experimentation with new business models make the need for sound financial management in coming years even more vital.

No one would deny that business and accounting professionals need to keep abreast of the complexities of media financial management, but this book is not intended for them. Instead, it is intended for the people who have other job descriptions, but who nonetheless have a stake in the successful running of a media business. The book also is intended for students of media management who need exposure to the essential principles and jargon of financial management. Business managers and accountants do not operate in a vacuum. They must rely on the competency and goodwill of other people within the organization. Essentially, every department within a station, network, or cable operation contributes to the overall financial health of the business. The purpose of this book is to introduce nonfinancial people to the fundamentals of financial management so they can communicate better with their financial colleagues and appreciate some of the inner workings of the business of media.

Your editors were asked by the Broadcast Cable Financial Management Association (BCFM) to oversee the creation of this book mainly because we are not financial pros, although we do have a solid background in broadcasting. Although each chapter would be written by a highly qualified expert, there was a concern that the authors might take too much for granted, and that editors possessing a similar financial background might not be sensitive to the learning needs of a nonexpert. Consequently, we were solicited to evaluate and modify the chapters not because of what we knew, but rather, because of what we did not know. During this process, we assumed that if we couldn't understand what an author was talking about, neither could an ordinary reader, and so we asked questions and made changes for the sake of clarity. We are both teachers and are familiar with the learning needs of students in an unknown arena.

We were flattered to be asked to perform this task, and we hope the reader will learn as much as we did.

Walter McDowell
Alan Batten

Introduction to the Second Edition

Although it may sound like hyperbole, the world of broadcast and cable finance has changed dramatically since BCFM Press published the first edition of this book in 1994. Then the Internet was a curiosity, and very few people had email—if they did, it was via a slow dial-up connection (2400 baud was considered fast!). Fax machines were considered cutting-edge technology; cell phones were very expensive, and their weight was measured in pounds, not ounces. A cable system with 100 analog channels was state of the art. Radio, TV, and cable ownership was disbursed among many, many smaller companies. Satellite television was just being launched, as were digital cable tiers. There was no such thing as high-definition broadcast television in the United States; nor was there satellite radio, HD radio, or podcasts. And, on the financial side, scandals such as those at Enron, WorldCom, and others, and the subsequent laws intended to restore investor confidence—primarily the Sarbanes-Oxley Act of 2002—had not been contemplated.

The original book was also noticeably short of cable content because the majority of the contributors came from the broadcasting side of the media business. What cable issues it did address focused primarily on cable programming (applicable to channels such as ESPN or CNN), and overlooked the unique financial considerations of cable operating companies—the companies that provide the packages of programming (and now data and telephony services) to consumers.

Despite these shortcomings, the Broadcast Cable Financial Management Association (BCFM) continues to get requests for the original book because there is no other source that explains the basics of broadcast and cable financial management for both the industries' nonfinancial managers and to financial personnel new to the industries. Despite its obviously dated contents, the original is required reading in several broadcast companies (followed, of course by, "Here's what's changed since that book was written . . ."). It has been used as a secondary college text as recently as 2004.

Working with Focal Press has allowed BCFM to produce a second edition that is a significant upgrade from its predecessor. Not only does this edition include information that is more complete and more current, with Focal's help, we have improved both the editing and the graphics. In addition, we have added an index and a glossary—very valuable additions because language in the media industry is riddled with acronyms, jargon, and familiar words that have unique meanings when used in the context of media and/or media finance.

This edition is organized in two parts. Part I covers the fundamentals: an introduction to the industry; the role of the financial manager; an overview of financial statements; key sources of revenue, followed by important information about industry expenses; an overview of the financial systems used in broadcasting and cable (including several that are unique to the industry or industries); a discussion of capital assets; a look at key performance metrics for the industries; and an overview of the extremely important issue of cash flow. Part II gives an overview of specialized areas that are nevertheless important to nonfinancial managers and those interested in understanding the electronic media business. These include Sarbanes-Oxley and internal controls; planning and budgeting; administration of credit and collection—an interesting topic because the industry sells a perishable product (advertising time and/or media transmissions to consumers), and generally tries to collect for that product after it has been consumed; trade and barter transactions; music license and syndication fees; taxation; and the financial and accounting considerations when media ownership changes hands.

The rewrite of this book is one of the tangible results of the work done by the BCFM Strategic Planning Committee to position the Association to meet members' needs both now and in the future. A not-for-profit educational association, BCFM's mission is:

To be the premier source of education, networking, information, and signature products to meet the diverse needs of financial and business professionals in the broadcast, cable, and electronic media industries.

This book would not be possible without the leadership of the project's Co-chairs—Leslie Hartmann and Timothy Pecaro. As 2006 BCFM Chair, Leslie Hartmann had the vision and the drive to make this rewrite a priority, and she committed to heading up the project when her Board term ended. Working with Timothy Pecaro, she developed the outline for the new book, drafted volunteers to write each chapter, and reviewed each submission to ensure that it covered the subject thoroughly and at a level easily understood by our target audience: nonfinancial media professionals; financial professionals new to electronic media; and academics—both professors and students concentrating in media management.

The volunteers who contributed their time and knowledge to this volume are all experts in their field. They are among the most senior financial persons in broadcasting and cable. This book simply would not have been possible without their contributions.

We were also fortunate to have Walter McDowell, Ph.D., and Alan Batten as our editors. They are experienced writers and editors, and, most importantly, they understand the business of electronic media. Before entering academia, Professor McDowell spent more than two decades in commercial television, including station management positions in promotion, programming, and creative services. After earning his doctorate in Mass Communications from the University of Florida, and teaching for several years at Southern Illinois University, he joined the faculty of the School of Communication at the University of Miami in 2001. In addition to teaching various media management courses, Professor McDowell has published media branding studies in several academic publications, including the *Journal of Media Economics*, and the *International Journal on Media Management*. A nationally known consultant, he has also coauthored two books published by the National Association of Broadcasters (NAB) and Focal Press: *Branding TV: Principles and Practices* (with Alan Batten) and *Troubleshooting Audience Research*.

Alan Batten graduated from the University of Maryland (College Park) and undertook postgraduate work at Boston University's College of Communications. He has been active in the broadcast industry since 1971, when he had the opportunity to lead the marketing efforts of several stations affiliated with NBC, ABC, CBS, FOX, and PBS. Along the way, he served as President of the Broadcast Promotion and Marketing Executives (now known as PROMAX). Batten has received countless industry awards and has presented numerous international lectures on various aspects of broadcast marketing. In 1991, he founded ABCommunications to provide consulting services for marketing challenges. ABCommunications counts as clients such industry giants as Sinclair Broadcast Group, Universal Pictures, The United States Postal Service, and Raycom Sports.

Finally, we owe a debt of gratitude to those who went before us. Without the first edition of this book, there simply would be no second edition. Although there have been significant changes in the world of broadcast and cable finance, other areas—such as financial statements, trade and barter, and taxation, among others—have changed very little over the years. Wherever possible, our authors used what was written in the first edition as the starting point for their contributions to this edition.

Any mistakes in this book are solely the responsibility of BCFM. In addition, because the world of media and media finance is changing so quickly, we encourage you to address questions about topics not covered in this book to us at BCFM. Contact information is available on the Association's web site—www.bcfm.com.

Part I

1 Introduction to Broadcasting and Cable

Timothy Pecaro and John S. Sanders

On the face of it, broadcasting and cable appear to be relatively simple industries—the first supported primarily by advertising, and the second by subscriber fees. Neither has inventory in the traditional sense, eliminating the need for storage facilities and inventory-tracking systems associated with traditional businesses.[1]

However, both industries have unique and complex financial issues that are dealt with in substantial detail in the body of this publication. These issues have grown even more challenging as the industries have evolved due to increased competition and the implementation of new technologies. As an introduction, this chapter provides an operating background and abbreviated history of each industry segment, and then discusses the regulatory environment in which these businesses operate. It is important to understand this environment because it impacts all the financial issues and systems discussed in this book.

General Industry Structure

Although the fundamental science that underlies broadcasting and wireless telecommunications was developed in the late 1800s and early 1900s, a true broadcasting industry did not develop in the United States until the 1920s, when radio stations licensed by the U.S. government began regularly scheduled broadcasts. Television broadcasts began in the late 1940s, and television ownership became widespread in the 1950s. The cable industry was born at that time, originally to broadcast signals to communities that could not receive over-the-air signals as a result of terrain or distance, a far cry from the telecommunications conglomerates, such as Time Warner and Comcast, that define the industry today.

---

1. Cable systems do maintain construction inventories and subscriber-access equipment.

The broadcasting industry in the United States is unique in the world because the government has played the role of a relatively detached overseer of the airwaves, and not an actual broadcaster. In the United Kingdom, for example, the British Broadcasting Corporation (BBC) was (and is) a government agency that not only regulated broadcasting, but was also responsible for producing and airing programming.

In contrast, the regime of broadcasting in the United States, from day one, was based on the premise of private ownership and management. The primary regulatory agency, the Federal Communications Commission (FCC), is responsible for technical matters such as allocating electromagnetic spectrum for uses ranging from broadcasting to cellular telephony.

The FCC was created by Congress in the Communications Act of 1934. Among the agency's responsibilities are "regulating interstate and foreign commerce in communication by wire and radio so as to make available, so far as possible, to all the people of the United States a rapid, efficient, nationwide, and worldwide wire and radio communications service . . . . " (The FCC says that "radio" in "its all-inclusive sense also applies to television.") The Communications Act charges the FCC to "make such regulations not inconsistent with law as it may deem necessary to prevent interference between stations and to carry out the provisions of [the] Act."

In exchange for a nominal license fee and a commitment to comply with the regulations made by the FCC, businesses are granted licenses for different portions of the spectrum, and as indicated above, the FCC ensures that they do not interfere with each other. Stations' commitments include such things as fostering public understanding of community issues by presenting programs and/or announcements about local issues (known as "localism," this commitment is usually fulfilled with local news programming); providing a specified number of hours of children's educational programming; limiting commercial advertising in programming targeted at children 12 and under; closed captioning of television programming to aid the hearing-impaired; participation in the Emergency Alert System; and many, many others.

Although the FCC does regulate content in some areas such as indecency, it does not dictate what a station will air or how a station is formatted as long as the station follows general licensing guidelines. Consistent with the First Amendment, and in contrast to many other countries, the FCC does not approve the editorial or news programming that a station or cable system can air. It is important to note that the FCC's ability to regulate content does not extend to cable programming networks such as MTV or ESPN because they do not use the broadcast airwaves to transmit their programming and because customers must pay a fee to receive them.

This relatively minimal level of government involvement sets the United States apart and has fueled the vibrancy of the media sector. Station owners develop programming based upon the preferences and needs of their audience. This has spawned extraordinarily popular programming evolving from *I Love Lucy* in the early days to *The Simpsons* today, a vibrant television-news industry, and entire networks that cater to the needs of specific sectors of the population, as evidenced by the large number of Spanish-language stations. Specialized cable channels have grown in response to demand for sports, history, and educational programming. Although controversies are ongoing regarding indecency, children's programming, and the like, decisions regarding broadcasting and cable programming are made in the marketplace, not in a government agency. Similarly, with some exceptions for public stations, broadcasting and cable are financed by selling services and raising capital in the markets, not by taxation or government assessments.

Broadcasting

The commercial broadcasting industry in the United States consists of approximately 600 VHF (very high frequency) and 800 UHF (ultrahigh frequency) television stations, along with approximately 5,000 AM (amplitude modulation) and 6,500 FM (frequency modulation) radio stations. VHF stations broadcast on Channels 2 through 13, and UHF stations broadcast on Channels 14 and above. The channels allocated for VHF service are in the 30 MHz (megahertz) to 300 MHz range, whereas UHF channels are in the 300 MHz to 3,000 MHz range. Due to the higher frequency, UHF signals are generally inferior to VHF signals and require significantly more electricity. Similarly, FM radio stations operate in the 88 MHz to 108 MHz range, and are generally associated with better fidelity than AM stations, which broadcast in the 535 kHz (kilohertz) to 1,705 kHz range. This difference in fidelity explains why the FM band has a higher proportion of musical programming, and the AM band has a higher proportion of news/talk programming.

There are additional stations that operate on noncommercial licenses or are licensed as low-power stations, translators, or boosters. In all, almost 28,000 licenses have been granted to operate broadcasting facilities in the United States. The FCC has granted each station a license that provides the primary authorization to operate. The license also stipulates certain technical parameters such as location, antenna height, and transmitting power. Station owners can petition the FCC in an open process if they wish to change a tower location or other technical parameters; entities that object to the change can then participate in the process.

With the advent of digital technology, the differences between the bands are diminishing. Radio stations are already broadcasting digital high definition (HD) programming, which permits them to multicast several channels of CD-quality programming on the spectrum allocation that had previously accommodated only one analog channel. Similarly, in February of 2009, television stations will be required to surrender their old analog channels to the Federal Communications Commission and broadcast exclusively on their new digital spectrum allocations.

It is estimated that over 600 million radio receivers are in use by the American public. Approximately 98 percent of all households in the United States are equipped with television receivers; 75 percent of all households own more than one television set.

Under the regulatory regime in the United States, viewers and listeners do not pay fees to the government in exchange for the right to receive programming. These stations provide entertainment, news, music, and other forms of programming to the public free of charge. In order to cover the costs of operation, commercial stations sell advertising time to local and national businesses, government agencies, and political organizations that are seeking to deliver information to the general public. Over time, though, they have also developed ancillary sources of revenues such as production fees, program syndication, tower space rentals, event sponsorships, Internet sites, and the like. More recently, television stations have begun to receive retransmission fees for the carriage of their programming on cable systems. With the advent of digital technology, other sources of revenue from multicasting and the provision of additional services such as data transmission may create more demands on financial systems.

Noncommercial stations cannot accept advertising in the traditional sense, although they generate revenue through sponsorships, contributions, merchandise sales, and government funding.

The link between audience size and advertising revenues is fundamental to the broadcasting industry. Broadcasters constantly seek to provide programming that will develop the widest appeal among radio listeners and television viewers. The more effectively the broadcaster is able to meet the preferences of the public, the larger the station's audience will be. The larger the audience that a station can offer to advertisers, the more advertisers will be willing to pay for time on the station. This relationship between audience size and advertising revenues is axiomatic in the broadcasting industry, and is the primary determinant of success or failure among station operators.

The quality of the audience that a station delivers is also an important factor in driving revenues. Services such as Arbitron and Nielsen provide detailed information on gender, age, income, ethnicity, and other factors that may make certain programs more attractive to advertisers of certain products than to others.

In recent years, the broadcasting industry has become increasingly competitive. The FCC has issued additional licenses for radio and television stations in almost every market in the country. Moreover, traditional broadcast operators have come under increasing pressure from satellite-distributed program services, cable television systems, compact discs (CDs), digital video discs (DVDs), portable music devices (iPods), Internet businesses, and other competing technologies. "Fragmentation" has become a buzzword in the media as a revenue pie that was once dominated by three major television networks—ABC (American Broadcasting Companies, now known as ABC, Inc.), CBS (Columbia Broadcasting System, now known as CBS Broadcasting Inc.), and NBC (National Broadcasting Company, now known as NBC Universal, Inc.)—has come to be divided among a growing number of outlets and technologies.

In order to build the largest audience share possible, stations invest heavily in tangible assets, such as tall towers and powerful transmitters, and intangible assets, such as on-air talent, broadcast rights, and syndicated programming agreements. Similarly, investments in equipment and intangible assets, such as managerial talent, may be oriented toward controlling costs and increasing profitability.

The importance of intangible assets is another factor that makes broadcasting unique. At a typical television or radio business, the preponderance of value will typically lie in intangible assets such as FCC licenses, advertising contracts, talent contracts, programming rights, and the like. Not surprisingly, industry parlance for the product sold by these industries, "airtime," describes something that can be extremely valuable, but cannot be seen or touched. Measuring, monitoring, and reporting about these intangible assets create unique challenges for financial reporting.

Ownership

In the Telecommunications Act of 1996, the FCC relaxed many of the rules regarding broadcast station ownership. This marked an important chapter in a process of deregulation that has dramatically changed the nature of the broadcasting industry. In the early days of the industry, for example, no single company could own more than 12 radio stations.

Because of the act, groups that had been limited to ownership of no more than a few dozen stations could now own hundreds of stations—and potentially could serve almost every major market in the country. The ownership limits within each market were also relaxed. For television, "duopoly" ownership was allowed, whereby an existing owner could acquire a second station in the same market as long as only one of the stations was among the top four stations in the market. The rules for radio were more complex, and were based upon market size and the number of competing stations within a market, including noncommercial stations. In the largest markets, an owner can own as many as eight stations, of which five can be in the same service (AM or FM). During this time period, many broadcasters grew rapidly; Clear Channel Communications, for example, grew to own over 1,200 radio stations. Increased demand during this period made FCC licenses much more valuable in the marketplace; in fact, licenses doubled and even tripled in value as companies competed to acquire additional stations.

In 2002, as an outgrowth of its congressionally mandated biennial ownership review, the FCC began a proceeding to review all of its ownership rules affecting broadcasting. Its intent was to relax ownership rules even further. On June 2, 2003, the FCC adopted new rules governing local and national television ownership; local radio ownership; and local cross-ownership of radio stations, television stations, and daily newspapers. Before the new rules took effect, however, the United States Court of Appeals for the Third Circuit imposed a stay on their effectiveness pending review. Although a detailed treatment of the ownership regulations is beyond the scope of this chapter, it should be noted that many of the ownership limits are still in a state of flux based upon judicial and FCC review.

Partially as a result of the regulatory uncertainty regarding ownership limits, and partially because of the downturn that hit the technology and media sectors after the year 2000, the pace of consolidation slowed. However, a significant number of television, radio, and cable businesses continue to change hands each year, and placing these businesses on the books of the acquirer is one of the most important financial functions in the industry.

It is in this marketplace, one defined by a strong relationship between audience size and revenues on one hand, and increasing competition on the other, that the broadcasting industry operates.

Cable

The cable television (CATV) industry developed in the late 1940s in order to provide television service to communities in rural Pennsylvania that were too isolated to receive over-the-air television broadcasts. The first systems consisted of a simple antenna placed on a tall hill that could receive television stations from distant markets. The cable system owner ran cables from this antenna location to households throughout the community, sometimes for free, in order to sell television sets locally. Like broadcasting, the industry has grown and diversified to provide a broad range of educational, entertainment, cultural, and sports programming to large urban areas and rural communities alike. These systems now provide telephony, high speed Internet, and business support services as well.

According to data from the National Cable and Telecommunications Association (NCTA), the cable industry in the United States consists of approximately 11,800 operating systems serving over 34,000 communities throughout the country. In addition, approximately 100 additional cable television franchises have been approved but have yet to be constructed.

The cable industry now serves almost 67 million basic subscribers, representing a 59 percent penetration of the approximately 113 million television households nationwide. Approximately 77 percent of basic subscribers also subscribe to a premium tier of service. Like broadcasting, cable television plays a significant role in the U.S. economy. For example, the NCTA says that cable systems spent $12.4 billion for capital and paid $2.8 million in franchise fees to local municipalities in 2006 alone.

Each system has been granted a franchise by its local municipal government or, more recently, by a state franchising authority. Municipal franchisees generally include guarantees that the cable operator will make expensive investments in local employment, local programming, and system technical design. The efforts of competitors from the telephone industry have resulted in statewide franchises that allow a competitive system to be built without approval from local municipalities.

The construction of a cable television system is extremely capital intensive. The cost of installing aerial cable is often the single largest investment made by a cable television system operator. Underground cable television installation is even more expensive, when considered on a per-mile basis. Additionally, investments must be made in headend facilities, satellite-receiving equipment, call centers, installation and service vehicles, warehouse and office facilities, and subscriber equipment such as converter units, which ultimately deliver cable television services to households.

Numerous changes have occurred in the development of cable television technology. Original systems used vacuum tube electronics and provided only a few off-air channels to subscribers. Companies have had to "rebuild" distribution plants over the years, installing cable with greater signal capacity and increasing amounts of fiber-optic cable and replacing the old vacuum tube technology several times over. Modern systems are capable of providing hundreds of channels of service, including satellite signals and locally originated programs. These systems use solid-state amplifiers and addressable converter equipment to control subscriber service levels.

Cable television systems provide movies, entertainment, news, music, and other forms of programming to the public. The cable operator must pay a fee, usually calculated on a per-subscriber basis, to program suppliers. These fees may either be determined on a fixed basis or calculated as a percentage of system revenues.

In order to cover the costs of operation, systems sell "basic" services such as local television signals, local origination programs, and some satellite services for a fixed monthly fee to all subscribers. Customers also have the option to subscribe to additional "premium," or "pay," services—such as HBO (Home Box Office) or Showtime—which offer movies, sports, entertainment, and original programming. Additional programming can be purchased in packages called "tiers," which can include additional news and information, specific sports programming, foreign language programming, high definition channels, and/or other types of programming networks.

Given the substantial fixed costs resulting from the capital requirements of the business, as well as high programming costs, cable operators seek to maximize system penetration. Two types of system penetration are of paramount importance in the industry.

The first is basic penetration, which is a measure of the number of homes subscribing to cable television as a proportion of the homes that are passed by cable. If 600 homes subscribed to cable service in a community of 1,000 homes, basic penetration would be 60 percent.

The second important measure is pay penetration, which gauges the popularity of pay services among those households that subscribe to basic cable service. If each of the 600 cable households in the example subscribed to 2 pay services, pay penetration would be 200 percent.

The linkage between basic penetration, pay penetration, and customer development is fundamental to the cable industry. Operators constantly seek to provide programming and services that will develop the widest appeal among local households. The more effectively the cable operator is able to meet the preferences of the public, the larger the system's subscriber base will be. This relationship between subscribers and revenues is axiomatic in the cable industry, and is the primary determinant of success or failure among system operators.

As is the case with broadcasting, the cable industry relies heavily upon intangible assets, although the capital asset requirements for a cable system are also tremendous. Multiple headend facilities—which contain satellite downlink and cable and fiber-optic transmission facilities—are required, as well as miles of buried and aerial cable, and additional assets at each household location. Fleets of installation and repair vehicles are also necessary. However, the value of these assets is often eclipsed by intangible assets, such as franchise agreements (which authorize the right to provide cable service in municipalities) and the base of paying subscribers.

Although cable program networks collectively have made significant inroads, traditional broadcast television stations continue to be the mainstay of television viewing in the United States. Even at their best, cable network viewing rarely exceeds 5 percent of U.S. households, and usually falls in the low single digits, whereas popular programming on any of the major broadcast networks can routinely fall in the 10 to 15 percent range.

Cable Regulation

Cable television franchises were initially awarded by municipalities based upon a competitive application process. Cable operators received an exclusive franchise, and in exchange agreed to pay a fee, usually calculated as a percentage of revenues. They also agreed to adhere to certain standards regarding buildout timetables, service quality, channel offerings, the provision of service to educational institutions, and the like. Similar to the broadcasting regulatory regime, the government leaves alone the management of the business, and theoretically steps in only when a franchisee runs afoul of the regulations.

In October 1992, following a period of significant deregulation that began in 1984, the cable television industry was placed under increased federal and local regulatory control as a result of public concerns regarding rates and service levels. Under the provisions of the Cable Television Consumer Protection and Competition Act of 1992 ("the Cable Act"), the FCC was directed to develop policies and regulations that would address current cable television rates, future rate increases, competition, franchising, broadcast station carriage, and service standards. Several subsequent regulations stipulated price rollbacks and price controls affecting virtually all cable systems. The financial pressure in these regulations also impacted cable programming networks working to be added (or "launched") on cable systems.

Since 1996, the cable industry has undergone significant consolidation, with the largest cable companies (MSOs, or multiple system operators) acquiring many of the medium- and smaller-sized operators. The strategy of these operators was to consolidate systems into regional clusters. These clustered systems could be operated more efficiently and could utilize centralized technical facilities (headends) to distribute programming throughout the metropolitan service area. The largest cluster, Cablevision's New York City system, serves more than 3 million subscribers. Large clusters in metropolitan markets such as Boston; Washington, D.C.; and parts of Los Angeles each serve more than 1 million subscribers. Consolidation also allowed the system to begin selling advertising on its cable channels to create an additional revenue source.

These companies invested heavily in upgrading their distribution systems with fiber-optic cables to provide substantially increased channel capacities and the ability to provide advanced services such as digital cable tiers, high speed Internet access, and telephone service. As a result of the anticipated cash flow from these new services, benchmark prices for cable systems increased dramatically. Because of the numerous revenue streams—including video subscriber fees from basic service, digital tiers, and HDTV, advertising revenue, high speed Internet access, and telephone service—the revenues of clustered cable systems can dwarf those of the most successful local television stations. However, the expenses can also be extraordinary to construct and operate the systems.

Cable programming networks have also seen significant consolidation since the late 1990s. Increased consolidation on the cable system side of the business gave cable operators significant power in carriage discussions with independent networks. In response, cable networks with strong brand identity have either acquired lesser-known networks or launched new channels on their own, using the strength of the popular networks as leverage when negotiating carriage agreements for lesser-known networks.

In Conclusion

Although seemingly mundane, finance can be as interesting as the industry that it serves. Fortunately, the broadcasting and cable industries continue to evolve and flourish, spurred not just by the government, but by the dynamism of competition and private-sector ingenuity. This environment will provide ample stimulation and challenge for those involved in finance and accounting. These functions provide the vital controls that ensure the success of the business, and also provide to nonfinancial managers information that facilitates effective planning and decision making.

To support and track the progress of a business, there must be proper accounting for income, expenses, assets, liabilities, and equity. As the entertainment and communications businesses have grown to the forefront of American society, the role of finance and its systems has adapted, grown, and improved to meet the challenge. The years ahead will be exciting ones indeed for financial professionals in the broadcasting and cable fields.

2 The Role of the Financial Manager and Other Personnel

Leslie Hartmann

From an audience perspective, the business of media must seem like mostly fun and games, but in fact, it is a serious endeavor that demands intelligent, dedicated professionals keeping track of what often are enormous sums of money. Behind every financial statement is a person working within an organization. Leslie Hartmann, Regional Director of Business Analysis for Entercom Communications and a former Chairperson of the BCFM Executive Board, explains the role of a financial manager and that of several other vital employees who help media companies avoid financial chaos.

Introduction

This book will provide you with insight into many of the unique and relevant areas of finance in the broadcast and cable industry. However, important as systems, regulations, and the financial processes are, it is the people who pull all of these things together to make an organization successful. This chapter provides an overview of financial organizational structure and a brief description of the various roles and responsibilities of the financial personnel and the skills needed to perform those functions.

Centralized versus Decentralized Companies

To describe a company as centralized usually implies that many of the accounting and finance functions have been *consolidated,* either regionally or at a single corporate office. In a centralized accounting environment, the strength and the number of financial personnel at the operating units is extremely limited compared to those of the corporate finance team. Of the many benefits associated with a centralized accounting environment, cost efficiency and quality consistency typically are the most compelling. A company's organizational structure often will contribute to this decision in that if a firm is relatively small, a centralized location may make it easier to control the company's activities. Even large corporations that are spread over many smaller markets in which the available talent pool often is limited may favor centralization. Functions that are usually centralized are financial reporting and accounts payable. In addition, many companies find efficiencies in cutting checks from a centralized location, thus maintaining tighter controls over cash and minimizing the number of checks cut to vendors. Some corporations with similar clientele across several markets will even centralize the credit function to avoid duplicating work.

In a decentralized accounting environment, individual divisions or local stations are responsible for maintaining many of the daily responsibilities, including much of the financial reporting. In contrast to the situation in a centralized environment, the corporate office typically maintains a small corporate finance staff, relying on the strength and the resources at the level of the lower operating unit. A major benefit of this type of structure is that the company can be more responsive to changing market conditions. Also, checks can be cut more quickly, and important information can be provided almost instantaneously to management. Another benefit of decentralization can be a higher level of accuracy when market financial managers each prepare financial statements or forecasts. Because they typically communicate with the department heads on a daily basis, they should be more knowledgeable about pending invoices that need to be addressed or incoming revenue that needs to be recognized. Most companies do not operate at one extreme or the other, but fall somewhere in between. Obviously, one can find both pros and cons to each structure, so there is no right or wrong approach, just a difference in corporate philosophy.

Size and Complexity

In addition to a company's structure, the size and complexity of the organization will often dictate the number of personnel employed in the various financial roles. For example, in smaller markets or entities operating in a decentralized environment, one person may perform many duties described in this chapter. For large organizations or clusters of stations or systems, there may be many employees handling the same functions. Some organizations have realized greater efficiencies by actually outsourcing some functions—such as accounts receivable, collections, and payroll—to specialized companies. Public corporations often require larger staffs, both to handle the additional reporting requirements demanded of public companies by the government, and to ensure strong internal controls, as will be discussed later in this chapter.

The Players

At a station or cluster of stations, one will almost always find a Financial Manager. The financial manager generally is hired by the station's General Manager or Corporate Finance Department, and is given dual reporting lines. The role of the financial manager continually evolves with changing regulations, and varies considerably depending on the type of company, its size, and its structure. Considered "the keeper of the cash," the financial manager's primary responsibility is to *protect the assets of the company*, which often includes the station's commercial inventory, trade and barter inventory, accounts receivable, personnel, property, equipment, and even the FCC license.

At a broadcast station, cable system, or even in a cluster of stations, the financial manager often has the title of Business Manager or Market Controller. Regardless of the working title, this person's primary role within the organization is managing the business office and all of its associated business functions. He or she is responsible for organizing the Finance Department and hiring the personnel to handle the responsibilities described below. In this managerial/custodial role, the business manager focuses primarily on current assets and liabilities on the balance sheet.

All financial managers need strong leadership and management skills, good analytical abilities, and a solid understanding of basic accounting and finance procedures. Unlike financial managers working in most other industries, there is no such thing as a "typical day" in the life of a broadcast or cable financial manager! These are dynamic industries, offering daily challenges for all who seek to make a living working in the media.

The financial manager handles the financial reporting of the entity; financial reports usually are prepared on a monthly basis, referred to as the month-end close. He or she must ensure that revenue and expenses are reported accurately and in accordance with Generally Accepted Accounting Principles (GAAP—standard guidelines that ensure that financial and accounting data have been assembled objectively and consistently). Reviewing and reconciling all balance sheet accounts, and preparing variance reports, the financial manager also attempts to explain why actual results may differ from budgeted or forecast income.

Recently, many financial managers have assumed the role of strategic business partner with their general manager. They work closely with all department heads to ensure accuracy in forecasting and recording revenue and expenses, and they advise management on strategies for improving efficiency and profitability. These strategies typically take the form of either (a) cutting costs or (b) developing new revenue streams. Financial managers often are responsible for reviewing economic conditions in their respective markets and developing the annual strategies and the operating budgets. Some financial managers provide return on investment (ROI) calculations for large promotions or capital investments. In addition, some managers work closely with their sales departments, assisting in commercial pricing and inventory control.

For companies that must report financials publicly, the role of the financial manager has become increasingly important. Previously regarded as just a "bean counter" or financial reporter, many financial managers today are responsible for ensuring strong internal controls, among which are the segregation of duties (e.g., the person who opens mail containing checks cannot be the person to make the bank deposit), ensuring proper authority (e.g., only specific positions have the authority to sign contracts), and the safeguarding of company assets. Although the financial manager has always been responsible for internal controls, firms have become more reliant on their professional skills in the face of the strict compliance requirements of the Sarbanes-Oxley Act, Section 404, which will be addressed in a later chapter in this book. These added responsibilities have enhanced the exposure, respect, and prestige of financial managers within public media corporations.

The following positions often report to the financial manager. In smaller organizations, the financial manager may even be responsible for performing many, if not all, of these functions.

The Credit Manager is responsible for establishing and administering the organization's credit policy, thus protecting the advertising inventory of a broadcast or cable company, market, or division. This person reviews credit history and conducts reference checks on potential advertisers to determine whether the ad buyer should be extended credit. Remember, in most cases, advertisers are not required to pay on an invoice until the entire schedule of commercials has aired. Consequently, the broadcasters or cable operator is giving short-term credit to the media buyer. In addition, a credit manager monitors the payment history of existing advertisers to determine if risk factors have changed, thus requiring a change to their credit terms. The credit manager is often involved in the sales-order approval process, and ensuring that funds are collected in advance for clients not granted credit. Within cable businesses, the credit manager performs similar functions in assessing the creditworthiness of prospective new cable clients and advertisers.

A Collections Manager will establish the organization's collections procedures and policies. This person protects the accounts receivable of the company, with the goal of limiting the amount written off to bad debt. In some organizations, the Account Executives are directly responsible for the maintenance and collections of their accounts. In these situations, the collections manager will provide assistance to the Sales Department by preparing statements, aging reports, and identifying problem accounts. Typically, the collections manager will tackle the more difficult delinquent accounts by making phone calls, sending past-due letters, and working with outside collection agencies when necessary. In many cases, the credit manager will personally handle the collections function. This individual must possess good communication and interpersonal skills. A successful collections manager must strike the delicate balance between enforcing a collection and still maintaining a viable business relationship with a valuable client.

Accounts receivable employees are responsible for the timely and accurate billing of advertisers, cable operators, and cable subscribers, as well as the overall management of the accounts receivable system. These people post payments to client and customer accounts, assist in resolving account discrepancies, and post billing adjustments and write-offs to ensure the accuracy of the aging report (the report shows the history of client payments and all open invoices) and statements. This function often requires strong communications and customer service skills.

The Purchasing Department is responsible for researching and obtaining bids from vendors, and placing orders for supplies and equipment. Accounts payable personnel are responsible primarily for ensuring that vendors are paid both accurately and within terms (that is, when due and not a moment before). These people protect the company's cash. The Accounts Payable Department will place and track purchase orders, ensure the accuracy of all of the invoices, get payment approval as required, code the invoices to the proper general ledger accounts, enter the invoices into an accounts payable system, and process the checks for the vendors. In addition, they maintain files of vendor contracts and track company leases. In many cases, they usually are also responsible for maintaining vendor W-9 files and processing the end-of-year 1099 reporting. Accounts payable personnel must have strong clerical and organizational skills. In a typical radio or television station, a single person or department handles both accounts payable and purchasing.

The Payroll Department is responsible for processing time sheets, salary adjustments, and all payroll deductions, including payroll taxes and employee benefits. They maintain payroll files and review or prepare employment tax reports—for example, SUI (state unemployment insurance), FUTA (Federal Unemployment Tax Act), and W-2's. Sometimes they are responsible for calculating and preparing sales-commission reports and calculating bonuses.

The Human Resource (HR) Manager is responsible for establishing and administering personnel policies. He or she maintains and sometimes negotiates employment contracts and manages all of the personnel files. This person administers the company's employee benefits program, oversees recruitment and hiring practices, handles employee disputes, and prepares any required EEOC (Equal Employment Opportunity Commission), OSHA (Occupational Safety and Health Administration), and state-mandated new-hire compliance reporting. At a station or cluster level, the payroll and HR managers are usually one and the same, but in a cable operation (which tends to be larger), they often are separate. In either case, they play a significant role in protecting the company's human resource assets. People working in these positions should have good communication skills and an in-depth knowledge of employment and personnel law.

In some broadcast and cable operations, the Finance Department may also be responsible for maintaining and managing trade inventory, managing the fixed assets, managing the public files, managing the traffic department, overseeing risk management, and even handling the front desk and other office personnel and office responsibilities, such as general office maintenance and ordering supplies. Some people say that finance gets to handle all of the jobs that no one else in the office wants to do. They are usually the first to arrive at work in the morning and the last to leave in the evening.

In addition to the above functions, cable programming companies generally have a Network Affiliate Finance (NAF) Department responsible for managing the billing and collection of license fees from companies such as cable operators and satellite companies, which distribute their program content. The single most important function of the Network Affiliate Finance group is affiliate billing and cash application. An affiliate is a cable system (or cable operating company) that carries a specific program network. That is, a local system (or MSO—multiple system operator) that provides A&E to its subscribers is technically an affiliate of A&E. This is similar to the situation in radio and TV where a radio or TV station can be an affiliate of a broadcast network, such as ABC, CBS, NBC, FOX, and so on. Depending upon the size of the network and the number of networks (sometimes also called "channels") under its control, the NAF group can be responsible for the billing, cash application, and accounts receivable management for hundreds of millions of dollars on an annual basis. The NAF group is responsible for managing relationships with hundreds of MSOs and representing thousands of cable systems. In situations where the programming company controls multiple channels, the number of relationships and dollar volumes can increase exponentially. Managing this large volume of transactions and the accompanying complexity requires (1) skilled staff, (2) a robust affiliate database (containing up-to-date information about each cable system carrying each channel and all contract terms for each affiliated system), and (3) best practices (that is, a consistent approach to managing the complex data inherent in the business).

Further complicating the job of NAF is that the bills sent to affiliates are based on estimates. The affiliates report actual monthly results with the payment. It should be no surprise that discrepancies are common. For this reason, traditional accounts receivable and credit and collection staffs are not the best candidates to fill these positions in an NAF organization; financial analysts are better suited for managing NAF responsibilities. The ability to understand complex contract terms coupled with advanced knowledge of database software used to manipulate and analyze large volumes of data are necessary skills for addressing and resolving payment issues.

Some finance functions are handled only on a corporate level. Although corporate finance responsibilities are very different from those found on the station or cable system level, they are quite similar to the finance activities in other industries and other corporate offices. These positions are typically focused on the long-term assets (value of property, equipment, and other capital assets expected to be usable for more than one year) and long-term liabilities (liabilities that extend beyond the current year) on the balance sheet. The positions typically found in the corporate offices include the Chief Financial Officer (CFO), Controller, Treasurer, Fixed Asset Manager, Tax Accountants, Financial Analysts, Internal Auditors, Risk Managers, and, in the case of publicly traded companies, SEC Accountants.

The corporate finance team works closely with the stations or divisions in developing policies regarding financial controls, financial reporting, and providing general support. They also are responsible for managing the company's cash and investments, consolidating financial statements, developing acquisition and financing strategies, developing organizational strategies, and setting and driving strategic goals for the organization.

In Conclusion

As we have seen in this chapter, trained and knowledgeable financial personnel are essential for achieving success. A thorough familiarity with the organizational structure, both corporate and local, is necessary in understanding the roles of the financial personnel, and the skills required of these professionals.

3 Understanding Financial Statements

Glenn Larkin as Updated by William Mangum[1]

To an untrained eye, looking at a typical financial statement can be a frustrating experience. What do all those numbers and terms mean? Can worthwhile information about the financial health and future of a media business be gleaned from this puzzling array of numbers and unfamiliar words? The answer, of course, is yes, given a little tutoring. William "Rick" Mangum and Glenn Larkin take away some of the mystery surrounding financial statements, and demonstrate how they can serve as valuable tools for making many management decisions. Rick Mangum is Vice President Broadcast Accounting for Clear Channel, one of the country's largest media companies. Glenn Larkin wrote the original version of this chapter while VP and Controller for Bonneville International Corporation.

Introduction

By analyzing financial statements, managers can make decisions based on fact rather than intuition or imperfect knowledge. Essentially, these statements constitute management's *road map* for monitoring the operating performance and financial health of a business. They also help managers to make intelligent decisions about cutting costs, discovering revenue-growth opportunities, improving productivity, and evaluating competition. A management team without the expertise to prepare financial statements or the ability to interpret financial statements is like a baseball team insisting that a shortstop play without a glove.

The accounting profession has developed standards of preparation of financial statements called Generally Accepted Accounting Principles (GAAP). These standards are driven by a number of conceptual guidelines that provide reporting consistency from company to company, industry to industry, and time period to time period. This chapter introduces these guidelines for financial statement preparation, and discusses the many uses to which the statements can be applied.

---

1. We would like to acknowledge the work done by Glenn Larkin for the 1994 version of this book, whole portions of which were again used here.

Financial Statement Uses

Each financial statement can serve the needs of many different users. A user may be defined as any entity, lender, regulatory agency, person, institution, or government entity that has a "need to know" and/or a "right to know" about the financial activities of a business. Financial statement users in a broadcast or cable setting may include station or company management (including department heads or project leaders), lending institutions, taxing authorities, investors, regulatory agencies, and certain employees. Each of these users requires specific information. For example, banks often need financial statements showing detailed liability information. Investors—interested in such metrics as earnings per share, dividend potential, and taxing authorities—require financial statements to substantiate income and deductions constituting the basis upon which taxes are assessed.

Internal use by management will likely represent the most common use of an entity's financial statements. Management relies on these reports to make daily operating decisions regarding the current and future course of the television or radio station or cable system. Very simply put, timely information given to management should result in accurate decisions leading to increased productivity and profitability. Such a variety of needs suggests that an accounting and financial system must be geared to fulfill all of the reporting requirements of a business. Further, preparing statements suitable for many users requires skills beyond simple accounting.

Financial Statement Preparation Conventions

To understand a financial statement, the reader must understand some of the conventions used in its preparation. *Financial statements basically function as a measurement medium.* Whether it is the measurement of a bottom-line result, of cash flow, or of total debt at year-end, each requires the use of certain GAAP guidelines. When these requirements are met properly, auditors examining the quality of these statements will issue what is called a clean opinion certified statement. The following constitutes a brief discussion of the more important GAAP conventions.

Cost Basis

The "cost" principle is used to value and record the expenses, inventories, broadcast rights, and other assets of a business at amounts that represent their purchase cost or some other acceptable cost measurement. This cost valuation brings "structure" to the measuring and reporting of activities, such as the portion of expended services or assets utilized in operating the business during a measurement period, or the portion that remains at the end of a time period. If an asset's cost exceeds its market value (the realizable value), the asset has lost value in that time period, and a charge for the amount of the decreased value must be recorded in the financial statements. In a broadcast or cable business, the "cost" convention applies to all asset and expense categories. It applies to the valuation of tangible assets, such as a transmitter or cable facility, as well as to an entity's intangible assets, such as talent contracts, FCC licenses, and cable franchise agreements.

Realizable Value

Another accounting guideline used in the preparation of financial statements is the "realizable value" convention. It is used in the measurement of revenues, sales, or gains of a business. The recording of revenues in the financial statements is measured by the value of the cash "realized" in return for the goods or services sold. For revenue to be recognized, several criteria must be met: (a) services must be rendered or goods received by the buyer, (b) there must be evidence of an arrangement with a fixed or determinable price, and (c) collection must be reasonably assured. In a broadcast or cable setting, this convention applies to commercial spots aired and cable programming provided to subscribers, as well as to all other types of customer sales.

When teamed together in the preparation of a financial statement, the "cost" and the "realizable value" conventions provide for meaningful measurement of net income or loss. Without these two rules, financial statements would include inconsistencies and unguided value judgments on the part of statement preparers and users.

An example of a misguided judgment would be the temptation to record revenues for signed but *unperformed* contracts, resulting in an improved bottom line that would encourage plans to float a public stock offering. Similarly, costs incurred in a transaction could be recorded at an amount less than the purchase price, temporarily increasing the bottom-line result. In this case, the reason may be a belief that future revenue potential justifies recording the lesser charge initially, and then carrying a higher portion of the costs into later time periods. The "cost" convention and the "realizable value" convention steer both preparers and readers of financial statements away from misrepresenting the true financial bottom line of a business.

Matching Principle

In the preparation of financial statements, the "matching" principle is used to define and measure those costs and expenses incurred in producing the revenues and resulting assets of an operation. This "matching" concept is particularly important in the preparation of the income statement. It requires that expenses, such as sales commissions, though perhaps not paid, be deducted from the revenues in the period the "revenues" are generated, in order to provide a meaningful measure of the *true return* at the bottom line. Hence, at the end of a measurement period, many unpaid expenses need to be accrued.

Where inventories are involved, this matching principle becomes even more critical because inventories associated with revenues realized in a financial statement must also be clearly reflected as a cost deduction in calculating net income. A simple example of this might be 2,000 DVDs sold at $25 each for a total of $50,000. Having been produced and held in inventory at a cost of $15 each for a total of $30,000, the gross profit on these video sales shown on the income statement should be the net of the two, or $20,000.

Similarly, though taxes are reported and paid according to a predefined timetable outlined by the tax authority, within the financial statements, taxes must be accrued and "matched" against the revenue or income to which they relate.

Conservatism

The GAAP "conservatism" principle requires that when it is difficult to evaluate the benefit of an expense or when an expense has questionable continuing value, that expense must be reported as a deduction early in the accounting process. This is to avoid carrying such costs on the financial statements as if the expenses were still contributing value to the business. Certain expenses may need to be accrued at the end of an accounting period to bring conservatism to the bottom line. Similarly, revenues and gains cannot be recorded until a bona fide revenue transaction is complete and any necessary contracts are signed to document that there will be "realizable value" received by the business. The conservatism convention avoids overstating profit and the value of assets, or understating losses and liabilities—all of which can negatively impact the reported financial status of an operation. For example, conservatism within a broadcast or cable setting suggests that purchased programming that is intended to be aired over several time periods, but whose value to any one period is not specifically known, should be taken as an expense in the financial statements on an accelerated timetable (i.e., in the earlier accounting periods) to avoid overstating the value of the program.

Conservatism would also dictate the speedy write down of an unpopular program acquisition that has deteriorated value to the business—meaning poor ratings, and therefore disappointingly low commercial rates. An example of deterioration might be a syndicated talk show that has provided a reasonable return for a TV station (i.e., good ratings). After a better-rated program is scheduled against the talk show by a competitor, the first show's value experiences a decline because the station will no longer command the same spot rates and resulting sales. Unfortunately, the station may be forced to absorb a cost or charge to its income statement.

Materiality

The convention of "materiality" requires that determinations be made as to the level of detail and the dollar level of activity (or balances) that should be reported to facilitate meaningful decision making. What constitutes a "material" amount for one business may not be sufficient for another business. Materiality in financial reporting is defined by the Sarbanes-Oxley Act of 2002 in the following manner: "Items are considered material, regardless of size, if they involve an omission or misstatement of accounting information that, in light of surrounding circumstances, makes it probable that the judgment of a reasonable person relying on the information would be changed or influenced by the omission or misstatement." In other words, a material item is one that *if reported improperly*, could cause people to make a false judgment or the wrong decision.

Although there are other accounting "conventions," those discussed here represent the basic framework within which financial statements are prepared. These guidelines provide for comparability, consistency, and understanding of the financial statements, whether used internally by management or externally by other users. Without these governing "conventions," numbers would be dropped into the statements more or less at random, thus rendering the financial statements less useful.

Accounting Periods

As the accounting conventions discussed above are applied to the business activities of an entity, a decision is required regarding the time period to be measured. Like any report, financial statements require the designation of an appropriate time period for the measurement of profits, losses, and reporting date for assets, liabilities, and owners' equity. (These terms will be defined in detail later in this chapter.) In general, the income statement and the statement of cash flow measure activity *over a time period,* such as individual months, individual quarters, or a combined year. On the other hand, a statement of financial position reports balances of assets, liabilities, and equity at a *specific date*. Financial statements can be prepared and published as frequently as management or other users desire; however, the administrative costs, labor, and time associated with preparation need to be considered.

Many broadcast and cable operations, like other businesses, use quarters and calendar years as their reporting time frame. A fiscal year may be designated in lieu of the calendar-year approach. Sometimes more than one reporting time frame will be used for different financial statement users. Once selected for reporting, an annual period may be broken down into monthly or quarterly interim-reporting time segments, depending on the need for more-timely financial data. Such interim-period financial statements are used by management, and may be less precise or less detailed depending upon their planned use. Most state and federal tax authorities, as well as regulatory agencies, require annual financial statements.

Management's report to stockholders is likely to be prepared on an annual basis, and is generally audited before being published. Audited financial statements require a higher level of scrutiny in their underlying accounting procedures.

Once the accounting periods and cutoff dates have been established for financial statement preparation, the same time frames should be consistently used over subsequent reports when possible. The continued use of similar reporting periods facilitates comparability in the underlying data from period to period. This comparability allows better evaluation of the progress of the business. Certainly, if there is a justification for changing accounting periods and report dates, that situation can be accommodated, but the advantages of year-to-year comparability will be lost during the transition.

Financial statements must contain a clear heading that states the time period they cover, using language such as "For the 12 months ended December 31, 2xxx," or "As of December 31, 2xxx." If prior-period financials are included for comparative purposes (and they usually are), the heading will reflect the periods and dates presented.

Financial Statement Descriptions

Four principal financial statements typically are prepared and published together. They include:

1. The Statement of Operations (sometimes referred to as the Income Statement or the Statement of Profit and Loss, abbreviated to P&L)
2. The Statement of Financial Position (sometimes referred to as the Balance Sheet)

3. The Statement of Cash Flows
4. The Statement of Changes in Shareholders' Equity (or Owners' Equity)

Although each statement has a separate and distinct function, they are interrelated, and, as business activity occurs, the data move among them. The following section discusses the basic purpose of each statement in a broadcast context, together with an illustration of each.

The Statement of Operations

This financial statement measures the operating results of the cable system or broadcast station/market, and thereby reflects that entity's profitability. The term "bottom line" literally comes from the last line of this statement, which usually indicates net income or net loss. The P&L (Profit and Loss) Statement is the common term used for internal financial reporting. Operating managers typically are evaluated and compensated based on the financial performance of the station as indicated on the P&L (see Figure 3.1).

Revenue includes all the sales and other taxable income of the business, excluding gains on the sale of property and any extraordinary items (see Chapter 4).

Operating expenses are all the costs of doing business, including salaries, employee benefits, supplies, and services required to operate the station efficiently—and, management hopes, profitably (see Chapter 5).

Depreciation and amortization constitute a special category of operating expenses, representing a *noncash charge* to the period being measured of costs associated with *long-term* assets. Depreciation represents the allocation of the cost of the station's or system's fixed assets—such as its studios and equipment—over the useful life of those assets. In a similar way, amortization constitutes the "depreciation"—or write-down costs of doing business—associated with intangible assets (nonmonetary assets that cannot be seen, touched, or measured), exhibiting a limited life span of usefulness. Certain indefinitely lived intangibles (e.g., trademarks and FCC licenses) are not amortized. These assets are subject to possible impairment charges (determined by subtracting the asset's fair value from its book value) if the discounted cash flows are not adequate to recover the cost of indefinitely lived intangibles.

A P&L statement may be more useful when other data are included against which current operations can be compared. For instance, actual results are commonly compared to proposed budget expectations, and also to prior performance of the same month one year ago, referred to as year-to-date information. The monthly

Station–Sample	Year ended December 31, 2007	2006
Broadcasting Revenue		
National	$1,422,470	$1,644,011
Local	4,368,652	4,112,523
Political–National	0	12,375
Political–Local	31,655	9,491
Gross Revenue	$5,822,777	$5,778,400
Agency Commissions	(596,561)	(523,858)
Net Broadcasting Revenue	$5,226,216	$5,254,542
Nontraditional Revenue	36,557	135,730
Other Broadcast Revenue	179,294	195,391
Net Revenue	$5,442,067	$5,585,663
Operating Expenses		
News & Programming	$1,718,334	$1,734,902
Marketing & Promotion	230,443	231,608
Technical & Engineering	181,561	198,048
Sales	1,225,123	1,320,260
General & Administrative	1,678,615	1,597,848
Information/Interactive Technology	56,387	107,507
Total Operating Expenses	$5,090,463	$5,190,173
Station Operating Income	$351,604	$395,490
Depreciation & Amortization	253,705	246,114
Operating Profit (EBIT)	$97,899	$149,376
Interest	$22,623	$23,762
Income Taxes	19,579	29,870
Net Income	$55,697	$95,744

FIGURE 3.1 *Broadcast P&L used for internal purposes.*

information gives management a sense of what happened in the most recent operating period, whereas the year-to-date figures represent cumulative results since the last fiscal-year reporting date.

Other operating analyses might include the calculation of earnings per share, profit margin, or percentage variances of revenue and expenses as compared with budget or prior periods. Expenses are commonly broken down into fixed and variable portions to determine controls that can be exercised realistically over discretionary spending. As the name implies, variable expenses are those costs that tend to *vary* in relation to increases or decreases in revenues, such as sales commissions. A sales executive who exceeds his or her expected sales budget increases station revenue—but at the same time, he or she increases commission rewards (a station cost) for achieving such a high sales figure. Again, as the name implies, fixed expenses generally remain constant despite variations in revenues. Examples of fixed expenses include transmitter maintenance and leasing agreements.

In the sample station P&L exhibit (Figure 3.1), most of the variable costs are found in the sales categories, such as commissions earned by sales personnel and national sales-representative firms. Most other broadcast and cable programming costs are relatively fixed across different levels of revenue over time.

A cable operator's costs, in contrast to broadcast, will increase as subscribers are added. This includes hardware, contracted installers' fees, subscriber-based program fees to networks (based on the number of system subscribers), and franchise fees to the communities served. Some costs do not increase proportionately with subscriber growth—such as a base level of trunk, headend, and feeder-system costs that are incurred initially and represent fixed expenditures. Both fixed and variable expenses have a significant impact on the well-being of a business. For best results, these expenses should be monitored and managed separately.

Certain items such as income taxes and interest payments are typically included separately at the end of the income statement. This manner of presentation is informative for several reasons. First, it facilitates the calculation of a profit *before* interest and taxes, which provides a pure measure of the *efficiency* of a reporting unit. This is because interest and taxes typically cannot be controlled by the management of a media operation, and therefore should not interfere with performance evaluations. Profit *after* interest and taxes, however, represents the true "bottom line," which is a critical measure for investors, lenders, and operating personnel.

Statement of Financial Position

This financial statement reports the balances of the assets, liabilities, and owners' equity of the broadcast or cable entity. Hence, it is sometimes referred to as the "balance sheet." A simple formula is used in formatting the balance sheet, *Assets = Liabilities + Owners' Equity*, suggesting that the assets owned by the operation are either financed by debt (liabilities) or owned by the partners or stockholders. The value of the assets must, therefore, sum to the total of the liabilities and owners' equity. The statement of financial position (see Figure 3.2) describes the types of assets owned and their nondepreciated cost, as well as other key data such as the amount of debt imposed on the operation.

BALANCE SHEET–Sample Media Company

ASSETS
(in thousands)

	December 31, 2007	2006
CURRENT ASSETS		
Cash and cash equivalents	$11,400	$8,278
Accounts receivable, net of allowance	169,534	150,565
Prepaid expenses	12,284	11,445
Other current assets	26,614	27,829
Total Current Assets	$219,832	$198,117
PROPERTY, PLANT AND EQUIPMENT		
Land, building and improvements	$88,651	$85,324
Structures	360,165	332,732
Towers, transmitters, and studio equipment	87,240	86,606
Furniture and other equipment	55,709	59,656
Construction in progress	9,264	9,061
	$601,029	$573,379
Less accumulated depreciation	279,935	249,587
Net Property, Plant and Equipment	$321,094	$323,792
INTANGIBLE ASSETS		
Definite lived intangibles, net	$82,281	$48,079
Indefinite lived intangibles–licenses	458,754	430,728
Goodwill	744,985	706,836
Total Intangible Assets	$1,286,020	$1,185,643
Total Assets	$1,826,946	$1,707,552

LIABILITIES AND SHAREHOLDERS' EQUITY
(in thousands)

	December 31, 2007	2006
CURRENT LIABILITIES		
Accounts payable and accrued expenses	$89,304	$73,110
Current portion of long term debt	33,637	89,118
Deferred income	14,369	11,667
Other current liabilities	2,176	2,027
Total Current Liabilities	$139,486	$175,922

LONG TERM LIABILITIES		
Long-term debt	732,670	615,536
Deferred income taxes	74,081	53,363
Total Long Term Liabilities	$806,751	$668,899
Total Liabilities	$946,237	$844,821
SHAREHOLDERS' EQUITY		
Common Stock, par value $.10 per share	$4,939	$5,383
Additional paid-in capital	267,456	279,457
Retained earnings	608,314	577,891
Total Shareholders' Equity	$880,709	$862,731
Total Liabilities and Shareholders' Equity	$1,826,946	$1,707,552

See notes to financial statements

FIGURE 3.2 *Statement of Financial Position.*

The categories of current assets and current liabilities are those that are expected to be realized or paid within the following 12-month period. Current assets represent those assets expected to be collected or consumed by the operation of the business within the 12-month period, such as cash or accounts receivable. Current liabilities represent the amount of payables or debt due during the same 12-month period, such as accrued bonuses and accounts payable. A financial statement reader can get a quick view of the near-term *financial stability* of the operation by comparing current assets with current liabilities on the balance sheet release date.

Other assets and liabilities of the operation are generally reported as "long term," suggesting that their contribution to the operation will exceed 12 months. The individual classifications of assets and liabilities reported on the balance sheet are likely supported by numerous accounts in the general ledger, combining to the totals shown on the report. The general ledger is a summary of every transaction that occurs, and serves as the building blocks, or raw material, for all of the other financial statements described in this chapter. Typically, any material items or balances, whether assets or liabilities, short or long term, are broken out and identified separately on the balance sheet to add transparency regarding business activity.

As the media entity conducts business—such as selling airtime, incurring expenses, or buying capital equipment—the balances of the assets and liabilities consequently increase or decrease on the balance sheet. The net change of all such activity is reflected in the equity section. The *net change* in equity (exclusive of dividends or other capital transactions) from one period to the next is also the *net income* shown on the income statement. Assuming the enterprise is profitable, equity can be distributed to owners and shareholders in the form of dividends.

Reading the balance sheet usually starts with current assets followed by various asset classifications, and concluding with liabilities. In reviewing liabilities, readers commonly compare the duration and size of the liabilities to the assets available to pay down, or liquidate, the debt. *The net result minus liabilities from assets is the equity, or book value, of the business.* When reviewing the balance sheet, the reader seeks to identify whether there has been overall improvement or deterioration in the owners' interests. Of course, the various users of financial statements will place varying degrees of emphasis on different sections of the balance sheet. A lender, for example, might give significant attention to the liquidity (i.e., easily converted to cash) of certain assets that will be called upon to pay down a loan. A system or station manager, on the other hand, may be interested in the cash position or the assets against which he or she might borrow money. A prospective buyer of the operation might be interested in the future cash flow generating *potential* and the value of the long-term assets such as an FCC broadcast license or a prime real estate location.

Statement of Cash Flows

The statement of cash flows, as one might guess, reports the sources and uses of dollars coming into and going out of the business during the period specified. It also reports on the beginning and ending dollar balances for the period. This statement reveals how changes in balance sheet and income accounts affect cash. The analysis is typically broken down into three categories: (a) cash flows from operating activities, (b) cash flows from/to investing activities, and (c) cash flows from/to financing activities (see Figure 3.3). From a management perspective, this statement indicates the operation's ability to meet its short-term debts. As such, it is particularly useful to bankers, investors, and local management.

What is included under "operating activities," "investing activities," and "financing activities" may vary among types of business, but the distinctions are more a matter of the degree of emphasis placed on the transactions that drive revenues and profits. Cash flow from operating activities summarizes revenues and expenses from day-to-day operations (e.g., sales, salaries, rents). Investing activities constitute the major areas of operations and asset categories in which cash is committed to ultimately generate a return to owners. Financing activities are those used to fund/finance business operations; they include debt and equity.

When evaluating a cash flow statement, it is important to look at the sources of the company's cash. Are operating activities generating enough cash to provide

STATEMENTS OF CASH FLOWS–Sample Media Company
(in thousands)

	Year Ended December 31, 2007	2006
CASH FLOWS FROM OPERATING ACTIVITIES		
Net income (loss)	$68,876	$93,566
Reconciling Items:		
Depreciation	48,306	47,381
Amortization	15,075	15,419
Deferred taxes	201,855	38,174
Provision for doubtful accounts	3,593	3,528
(Gain) loss on sale of assets	(6,933)	(4,788)
Changes in operating assets and liabilities		
Decrease (increase) in accounts receivable	(20,234)	(2,531)
Decrease (increase) in prepaid expenses	(2,398)	1,538
Decrease (increase) in other current assets	274	4,304
Increase (decrease) in accounts payable and in accrued expenses	8,336	(3,470)
Increase (decrease) in deferred income	1,165	(1,838)
Net cash provided by operating activities	$317,915	$191,283
CASH FLOWS FROM INVESTING ACTIVITIES		
Purchases of property, plant and equipment	($35,049)	($32,572)
Proceeds from disposal of assets	10,032	10,200
Net cash provided by (used in) investing activities	($25,017)	($22,372)
CASH FLOWS FROM FINANCING ACTIVITIES		
Proceeds from long term debt	$7,839	$3,579
Payments on long term debt	(86,635)	(198,604)
Dividends paid	(38,277)	(34,332)
Payments for purchases of common shares	(137,146)	(107,020)
Net cash used in financing activities	($254,219)	($336,377)
Net increase (decrease) in cash and cash equivalents	$38,679	($167,466)
Cash and cash equivalents at beginning of year	8,278	175,744
Cash and cash equivalents at end of year	$46,957	$8,278
SUPPLEMENTAL DISCLOSURE		
Cash paid during the year for:		
Interest	$46,139	$43,038
Income taxes	–	19,372

FIGURE 3.3 *Statement of Cash Flows.*

continued growth (investment opportunity) or dividend payments? If the cash flow is generally originating from financing, what are the company's debt ratios (debt capital divided by total assets)? Is there buildup in accounts receivable and inventories, reflecting a slowdown in cash flows with which to liquidate payables and debt commitments?

The cash flow statement functions as a bridge between the P&L and the statement of financial position. All three financial statements have a critical role to play in the analysis and monitoring of business activity. A financial statement user should have a complete set of all three statements before commencing a review.

Statement of Changes in Shareholders' Equity

The Statement of Changes in Shareholders' Equity itemizes the changes in equity over the period covered—including investments by owners and other capital contributions, earnings for the period, and distributions to owners of earnings (dividends) or other capital. This statement is required only when there has been a change that

Media Company Inc.
CONDENSED STATEMENT OF CHANGES IN SHAREHOLDERS' EQUITY

	Year ended December 31, 2007	2006
Common Shares		
Number of shares, beginning of year	2,000	2,500
Shell acquisition, retirement of Media Company Inc. shares		(2,500)
Shell acquisition, issue new Media Company Inc. shares		2,000
Number of shares, end of period	2,000	2,000
Common Stock		
Balance–beginning of year	$15,000	$15,000
Balance–end of period	$15,000	$15,000
Additional Paid–in Capital		
Balance–beginning of year	$119,154	$70,000
Contribution of The Pension Fund Company Inc.		24,154
Capital Contribution		25,000
Balance–end of period	$119,154	$119,154

	Year ended December 31, 2007	2006
Accumulated Other–Comprehensive Income		
Balance–beginning of year	$1,054	$951
Net change in unrealized (depreciation) appreciation of investments deferred Federal tax benefits of $294 in 2007 and $80 in 2006.	(556)	103
Balance–end of period	$498	$1,054
Accumulated deficit		
Balance–beginning of year	($9,255)	($3,067)
Net loss	(2,520)	(6,188)
Balance–end of period	($11,775)	($9,255)
Total Shareholders' Equity	$122,877	$125,953

See notes to condensed financial statements.

FIGURE 3.4 *Statement of Changes in Shareholders' Equity.*

cannot be readily determined from other sources—that is, something other than earnings that affects the value of shareholder equity. Figure 3.4 provides examples of some such changes.

Financial Statement Footnotes

The grouping, summing, and subtracting of numbers in financial statements in and of itself sometimes is inadequate to present a complete picture of business activities. Consequently, parenthetical notations are often found within the statements themselves. Another option is to include some text—referred to as footnotes, or just notes—in addition to the financial statements. The notes must be read in conjunction with the basic financial statements, and should provide additional explanation of the many accounts of an enterprise. The financial statements will reference relevant footnotes at key points within the data disclosures. Examples of items typically found in footnotes include significant accounting policies; a breakdown of asset types included in the long-term-asset category; information on acquisitions and divestitures; investments; tax disclosure; long-term debt, commitments, and contingencies; and share-based compensation.

Footnotes typically clarify in more detail information revealed in the standard financial statements, such as descriptions of revenue recognition conventions or detail on debt instruments. Footnotes may also explain unique aspects of the financial statements. In fact, to a sophisticated interpreter of financial statements, the footnotes may be the most informative portion of the financial statements.

It should be noted that the majority of the statements discussed in this chapter are examples of those used internally. Publicly reported statements prepared to conform to public accounting requirements may be presented in a different format. The purpose of internal statements is to help management evaluate performance. Publicly reported statements are intended to help shareholders both evaluate their investments and compare investments across industries.

In Conclusion

No operator of a broadcast station or cable system can effectively manage operations without the aid of standardized financial statements prepared on a monthly, quarterly, or annual basis. Recognizing that many broadcast and cable professionals are not accounting experts, having qualified accounting personnel review these statements with these operational managers is a good idea. Once management understands the financial patterns and trends uncovered in these reports, financial statements will become important tools in recognizing problems, seizing opportunities, and, ultimately, earning a reasonable return on investment for owners and shareholders.

4 Revenue

Trila Bumstead, Joyce Lueders, and Fidel Quiralte[1]

Ultimately, business is about *money,* but recognizing how money is made and spent involves a basic understanding of financial concepts and jargon unique to broadcasting and cable. Together, these three authors unravel the complexities of *revenue* in this chapter, and *expenses* in the next chapter. Trila is the Executive Vice President and Chief Financial Officer of New Northwest Broadcasters LLC; Joyce is the Business/Program Manager of WFLA-TV in Tampa, Florida; and Fidel is Vice President of Finance and Controller of the Game Show Network.

Introduction

From the sale of products and services to customers, businesses generate revenue. This chapter discusses the revenue streams typically associated with the broadcast and cable industries, and explains how these industries account for those revenues. As the industries evolve with growing pressure from new competitive sources, new revenue sources continue to emerge, so the items in this chapter should not be considered a comprehensive list.

Revenue for broadcasters is generated primarily through the sale of local, regional, and national advertising on the local stations and their networks. The primary revenue generator for cable systems is subscriber fees (as of this writing, cable systems typically generate only 5 to 10 percent of their revenue from advertising). From a cable network programming perspective, the proportion of revenue acquired from cable advertising versus cable subscription varies, depending on the type of program content. For example, most premium services, such as HBO, have almost no advertising. The primary revenue generator in this case is a subscriber fee shared between the cable operator and the program content provider. On the other hand, highly successful program "basic" networks, such as ESPN, can command both significant per-subscriber fees from systems and impressive commercial rates from

---

1. The authors gratefully acknowledge Richard "Dick" Petty, SVP/Controller, Time Warner Cable, for his assistance with the portions of this chapter dealing with revenue from cable operations.

national advertisers. At the other end of the basic spectrum are less popular program networks that receive little or no subscriber compensation from cable systems, and struggle for advertising dollars.

The Broadcasting Model

Each station's local sales staff solicits advertising either directly from a local advertising client or indirectly through an advertising agency. Stations incur an agency commission based on gross revenue associated with the advertising dollars placed on the station that are secured through an advertising agency or specialized firm. Stations report revenue as gross revenue less agency commission to achieve net revenue. National advertising dollars are acquired by a station through the help of a rep firm, which serves essentially as a broker between the station and major advertising agencies representing national advertisers. The rep firm is compensated through a commission formula based on net revenue generated by that firm. Depending on the structure of the representation agreement between the station and the national rep firm, the commission paid to the rep may be reported as a sales expense or may be netted out against revenues in a manner similar to that used for agency commissions. (See Chapter 5 for more on selling expenses.)

Revenue is recognized when it is realized, or realizable *and* earned. Four basic criteria must be met to satisfy the realized (or realizable and earned) threshold in order to recognize revenue:

- There is persuasive evidence that an arrangement exists (i.e., client-approved order or an insertion order from an advertising agency).
- Delivery has occurred or services have been rendered (i.e., the commercials have aired, the concert has happened, etc.).
- The seller's price to the buyer is fixed or determinable (i.e., spot rate is determined, sponsorship amount is stated, etc.).
- Collectibility is reasonably assured (i.e., account is in good standing, credit application is approved, etc.).

For operational purposes, stations separate revenue into several categories to help managers evaluate performance, build budgets and projections, and explain variances. The most common classifications of advertising revenue for radio and TV stations and cable companies are listed and discussed in more detail below. They fall into four distinct categories: on-air advertising, nonspot revenue, emerging revenue, and cable-specific revenues.

On-Air Advertising (or Spot Revenue)

Unique to electronic media, the most valuable inventory is commercial inventory, or airtime. Unlike tangible goods, commercial inventory is very *limited,* in that audiences will tolerate only so many interruptions within program content. This inventory is also *perishable,* in that once the inventory time lapses, it can no longer be sold, similar to seats on an airline flight. Advertising clients purchase the commercial inventory to use as spot, or on-air, advertising. Spot advertising typically is sold according to varying lengths, such as :60s (60-second spots), :30s, :15s, and :10s. The commercial value of the spot is based on viewership, time of day, and, in the case of spots sold on cable systems, channel location (not all channels are available to all subscribers).

Spot rates are typically determined by sales management or general management based on complex calculations reflecting client demand, market competition, supply (available inventory), and seasonality.

For radio, audience ratings by day part influence the value of a spot. The more listeners (i.e., the higher the audience share) a station has, the more valuable a spot. This is often referred to as CPM (cost per thousand) or CPP (cost per [rating] point). Because audience numbers and time spent listening vary considerably, commercial rates for these day parts are not always the same. For example, commuters tend to spend more time listening to the radio in the morning and afternoon drive times, so those rates typically are the highest on a station.

For TV and cable, audience ratings differ by day part and by show, with prime time attracting the largest audiences, and consequently commanding the highest commercial rates on a station. As with radio, commercial rates often are calculated using CPP or CPM metrics. The actual audience size is merely an estimate or projection based on prior ratings performance for the day part or program. Once the purchased commercials have aired, the estimated CPP and estimated units delivered are used to create an invoice. Then, when the station or network receives the actual audience ratings, they are used to true up the actual delivered units and the final revenue. This may result in an adjustment to revenue if the delivered units were more or less than the units agreed upon. If the delivered units were less than agreed, revenue needs to be reduced, and a liability needs to be created. The liability is normally called ADU (Audience Deficiency Unit) liability. This liability will be reduced when "make-good" spots are aired to resolve the deficiency in audience delivered. A refund of the deficiency to the advertiser, otherwise called cash back, can also reduce the liability. By and large, finance professionals rarely (if ever) are notified about overdelivery of units. In the unlikely event that a media company expects an additional payment for schedule due to an overdelivery, the "conservatism principle" (see Chapter 3) prevents the company from recognizing the revenue before it is received.

Below are the various revenue types for spot revenue:

- **Local Direct Revenue**—This category of revenue is the traditional spot time sold to a local advertiser by in-house media sales representatives. These representatives are employees of the station.
- **Local Agency Revenue**—A local agency acts as a representative for a local advertiser, and usually places that advertiser's entire advertising budget with all media (radio, TV, newspaper, outdoor, and the like) in the marketplace. The agency earns money by withholding a percentage of the advertiser's *gross purchase* (typically 10 to 15 percent) as a fee, and the station receives the *net proceeds* after the *agency discount*. This discount is referred to as an agency commission.
- **Per-Inquiry or Direct-Response Revenue**—This revenue traditionally is included in the category of local direct revenue or agency revenue, but instead of an agreement based on CPM or CPP for the commercials to be aired, consideration and payment for the airtime is determined by the advertiser based on the *results* of the advertising—either by the number of inquiries (e.g., phone calls, emails, web site clicks, etc.), or predicated by the actual dollar sales that an air schedule generates.
- **Regional Revenue**—Revenue placed by agencies outside the local station's market for advertisers in the surrounding region is reported in this class.
- **National Revenue**—This advertising is purchased by large national advertisers and is placed by the station's national rep firm or directly with the network (either broadcast or cable).
    - What is a national rep firm? This kind of firm functions much like a local agency, but instead of representing the advertiser, they represent the station or cable system to larger national advertisers. These firms charge the stations (or systems) a fee similar to that for local agencies for their services. These fees are either paid as a sales expense or deducted from the net revenue form the advertisers' media purchases.
- **Network Revenue**—Stations may agree with network on-air content providers to air their programs in exchange for a fee or placement of the network's commercials during the program run. The revenue generated from this activity is

considered network revenue or network compensation. This category may also include ads placed in unsold inventory by a third-party company representing its own group of clients. This can occur when a station or system signs a contract to accept spots in unsold inventory with such a third-party company.

- **Trade Revenue**—Stations may agree with advertisers to accept goods or services instead of cash for advertising. In these cases, the revenue generated from the spot is classified as trade revenue, or barter revenue, for income statement purposes. As the station uses these goods or services, an expense is recognized to account for the usage.

Generally speaking, recognition of revenue for spot advertising described here occurs when a contract is approved by the client *and* the commercials air on the station, as outlined in the contract. As indicated above, invoices for commercial schedules are issued after the spots have run, so this is when the receivable is created.

Nonspot Revenue/Nontraditional Revenue (or NTR)

Until the 1990s, almost all broadcast station revenue was generated from spot advertising. As competition increased, however, stations had to develop new revenue streams, leveraging their vast audience in new ways. The following are some alternative sources of revenue.

- **Events and Concerts**—This category includes exhibitor revenues from job fairs and other station- or system-managed community events, ticket sales, and concessions sales.
- **Sponsorships**—Whereas advertisers can sponsor a special program for on-air content and potentially flow through local direct revenue, in many cases the most material types of sponsorships are related to special events such as concerts or special-event shows, which would place them in this NTR category. Also included in this category would be the sponsorship of a station van or other promotional item.
- **Merchandising**—Sales of station goods (T-shirts, bumper stickers, key chains, etc.) are categorized as merchandising.
- **Talent Fees**—This is a pass-through revenue account in which the employer recognizes (as expense) amounts paid to on-air talent for appearances, and then

is paid for those appearances by the advertiser. Because employers must contribute the employer taxes on these amounts, the expense related to this talent fee is recognized in production and programming expense, as discussed in the next chapter.

- **Auctions**—Although on-air auctions have gained in popularity over the past few years, many stations are moving their auctions to the web. In any case, the revenue generated from the sale of auctioned goods would be classified here. Stations and systems often procure those goods to sell at auction via trade.
- **Publication Revenue**—Some stations, systems, or channels may publish magazines or guides that include advertisements. This is especially common for sports stations and companies that have a strong niche audience (e.g., football season guides, magazines produced by channels targeted at children, magazines for do-it-yourselfers, etc.).
- **Rental Income**—Stations that own their broadcast tower or studio facilities and have other tenants (cellular telephone companies, utilities, or other broadcasters) that lease space would recognize that income as rental income.
- **Retransmission Consent**—FCC rules allow television broadcasters to negotiate with local cable systems for compensation in return for permission to transmit the television broadcast signal on the cable system. Although this is a source of revenue for the television broadcaster, it is an expense for the cable operator (see Chapter 5).
- **Syndication Fees**—Companies that have developed strong programming content and on-air personalities that are popular and compelling have been able to distribute (syndicate) these programs to other broadcast markets and charge the stations a syndication fee.

Typically, recognition of revenue for nonspot revenue described above occurs only when the event is completed. However, the timing of revenue recognition might be more complex if an event sponsorship fee is coupled with a commercial spot schedule. In this case, the sponsorship fee would be recognized once the event occurs, but the spot revenue would be recognized *as the spots air*. Sales managers will argue that sponsorship fees may be recognized prior to the event because of on-air mentions or "ticket-stop" appearances that are scheduled prior to the actual event. The question should be asked, "If the event is canceled for any reason, would the advertiser expect a refund of their sponsorship commitment?" If the answer is yes, then the revenue should not be recognized until the event actually takes place.

Revenue Adjustments

Adjustments to revenue are considered a contra revenue account, and result in an offset or reduction in advertising revenue. Revenue adjustments are typically caused because (a) a sales order was improperly entered into the system, (b) a spot aired out of schedule, or (c) incorrect copy aired for a spot. Some companies choose to anticipate sales adjustments based on historical records, and automatically subtract a percentage from expected revenue. Other companies simply book the adjustment at the time that the station is notified of the incorrect advertising.

Emerging Revenue Streams

At this writing, Internet revenue is the most rapidly growing source of advertising expenditures, coinciding with the exponential growth of Internet usage. In addition, stations and systems are capitalizing on their large audience base by moving their content to digital platforms. The potential revenue streams continue to expand. In addition to programs, media companies can also provide interactive video and data services. Emerging revenue streams may include:

- **Internet Revenue**—This is revenue associated with audio streaming, portals, banner advertising, embedded links, and new web sites.
- **Digital Broadcasts**—Broadcasters can now provide two or more high definition (HD) programs or even more channels (multicasts) simultaneously using standard definition (SD) technology. The number of programs a station can broadcast at the same time depends on the level of picture resolution in the programming stream (i.e., the more channels used, the lower the resolution).
- **Video on Demand (VOD) and Podcasts**—These are a significant area of new revenue for traditional media outlets such as TV and radio because consumers have demonstrated that they are willing to pay for the convenience of controlling when and where they access media content.

Recognition of revenue generated from these new revenue streams will depend on the service promised. For instance, if a station sells a Flash banner advertisement for six months at a set fee, that fee would be recognized over the six-month period. In another instance, if a minimum number of web site "visits," or "hits" to a specified site, are promised, then the revenue should not be recognized until that target is achieved. Because revenue generated from these areas is rapidly changing, personnel in charge of emerging revenue streams should consult the accounting manager or controller to discuss revenue recognition of specific transactions.

Cable-Specific Revenues

Cable Network-Specific Revenue Streams

- **Affiliate Fees**—Cable networks and their distributors—cable systems, MSOs (multiple-system operators), direct satellite providers, and so on—normally have contracts that specify the terms and length of the agreement. The contract also specifies the fees the distributor pays the networks for carriage of the signal. These fees are based on the average number of monthly subscribers. More-complex license fee agreements will include terms that are based upon specific carriage requirements, volume discounts, channel placement (channel number and/or channels next to the channel being discussed), penetration (percentage of potential customers who subscribe to the distributor's service), and packaging incentives.
- **Launch Support Fees**—A launch support fee is a payment by the cable program network to the distributor to help launch a new cable channel. It is also called a subscriber-acquisition fee. Launch Support is a contra revenue account. These fees are normally capitalized and amortized over the contractual term of the applicable distribution agreements as a reduction in subscriber fees revenue. If the amortization expense exceeds the revenue recognized on a per-distribution basis, the excess amortization is included as a component of cost of service. This situation occurs when a network is attempting to increase the subscriber base, and justifies the higher cost of launch support by increasing revenues from advertising to pay for the growth.

Cable System-Specific Revenue Streams

Cable systems have evolved from simple one-way distribution companies to diversified broadband companies offering video, voice, and data services to both residential and commercial customers. On the residential front, services are offered on a subscription basis as well as on demand or pay per view (PPV). Cable system-specific revenues may include:

- **Video Revenue**—Cable systems or MSOs normally have contracts with programmers that permit carriage of video services. Operators offer these services in *bundles* based on FCC rules for carriage. Several packages (called "tiers") of nonpremium channels are generally offered, with the tiers offering the most channels being the most expensive to purchase. Additional monthly fees are generated by premium channels such as HBO and Showtime. Cable also offers a wide variety of digital and high definition TV options as enhanced video service offerings. Monthly services typically are billed in advance. Subscribers to cable systems pay a fixed monthly rate based upon the level of service to which they subscribe (i.e., which package).

- **Pay per View (PPV)**—Some individual programs, movies, and events are sold to subscribers on a per-viewing basis. Charges are determined by the type of event and anticipated demand, with movies being the most popular purchases. Events such as concerts and sporting contests (e.g., wrestling or boxing) are sold based on contracts with promoters. Sports programming can also be sold as a "season ticket" for an extended series of professional events ranging from football to baseball to basketball to hockey over the course of a season. Note that this is also an opportunity for the cable operator to work with the local television broadcaster. See Nonspot Revenue/Nontraditional Revenue (or NTR) above.
- **High Speed Internet Service**—It is sold in varying packages based on the delivery speed and features offered.
- **Launch Support Fees**—From a cable operator perspective, the accounting is the mirror image of what is done for networks (see above). Cash is paid by the programmer, either up front or as the launch is achieved, and is recognized as deferred revenue (subject to normal current-versus-noncurrent considerations). Such credits are amortized (usually straight-line amortization) over the term of the contract as a "contra," or reduction of programming expense. To the extent that the support received exceeds the programming expense incurred, it is recognized as miscellaneous revenue.
- **Telephony Services**—Cable operators are offering both traditional land-based products and wireless telephony. The charges for local service, features (e.g., call waiting, voice mail, conferencing, etc.), and long distance are similar to those for traditional telephone companies.
- **Business Services**—Many cable operators also offer services for commercial accounts. Available products may include video services for public commercial establishments (e.g., bars or restaurants), Internet, and/or phone services targeted for business customers.

The primary differentiator for cable operator–provided services when compared to individual service providers is the bundled approach for all three products (i.e., video, data, and telephony) at a discounted price with integrated features and billing not typically available elsewhere.

In Conclusion

This chapter has focused on many of the revenue sources available in the broadcast and cable industries. As discussed above, the sources of revenues are changing dramatically. Regardless of the source of revenue, it is important to follow GAAP guidelines and to recognize revenues only when they have realizable value.

5 Expenses

Trila Bumstead, Joyce Lueders, and Fidel Quiralte[1]

Whether you are running a neighborhood lemonade stand or a major media corporation, *it takes money to make money* for most business ventures. The lemonade retailer needs to pay for lemons, sugar, glasses, and a table before the first customer strolls by. Similarly, a broadcast station or cable system needs to acquire all types of program content before the first audience member or potential subscriber will pay attention. After addressing the issues of revenue in Chapter 4, the same three authors now look at the "other side of the coin," and address the necessary information needed to monitor—and, more importantly, *control*—business expenses.

Introduction

Operational expenses are recognized on an accrual basis (not on a cash basis), and consequently expenses should be "matched" to the appropriate revenue and time period. In broadcast and cable, the expenses include variable expenses (such as sales commissions and programming fees), as well as fixed expenses (such as salaries and benefits, maintenance, marketing, IT, leases, and utilities). Radio stations typically enjoy a higher operating margin (operating income divided by net sales) than other broadcasting or cable entities due to low fixed costs and lower variable costs associated with revenue. Large-market VHF television stations (stations operating on channels 2–13) can also exhibit high levels of profitability because they can command more dollars from advertisers, whereas both their fixed and variable costs represent a lesser percentage of sales than do those of their smaller-market brethren.

In any case, expenses associated with broadcasting can vary dramatically depending on such factors as the cost of programming, market size, radio station format, and geographic location. However, most media operations experience the same

1. The authors gratefully acknowledge Richard "Dick" Petty, SVP/Controller, Time Warner Cable, for his assistance with the portions of this chapter dealing with revenue from cable operations.

type of expenses. These cost centers typically are organized into departments, such as Sales, News & Programming, Marketing & Promotions, Technical & Engineering, Information/Interactive Technology (IT), and General & Administrative (G&A).

Expenses common to all departments and not easily broken out or identified among departments are often combined within the G&A category. As outlined below, these expenses include employer-paid taxes—such as FICA (Federal Insurance Contributions Act, which includes Medicare and Social Security), FUTA (Federal Unemployment Tax Act), and SUI (state unemployment insurance)—as well as employee benefits, such as medical, dental, long- and short-term disability, workers' compensation, and life insurance. Other benefits common to company employees are vacation pay and employer contributions to 401(k) plans. Office and other supplies also fall into this category of common expenses that are not easily differentiated among departments; however, some companies insist that the individual departments should be accountable for these work-related expenses.

The reporting of trade expenses also varies significantly among companies. Some operations elect to have trade expensed by category or department based on the usage. Others prefer to compile all trade into single revenue and expense accounts. In some cases, both trade revenue and expense are separated from other revenues and expenses and are shown in a separate area of the financial statement. In other cases, companies choose to exclude trade expenses from the profit and loss statement. Trade expenses are not included in the calculation of cash flow.

Many broadcasting companies operate several stations in a single market, and many cable operators serve several community franchises in a single market. These groupings are referred to as "clusters." Thought should be given to the allocation methodology of shared expenses such as salaries for General & Administrative Department staff or office rent. Some companies allocate these costs as an equal percentage distributed over all stations/systems. Other expenses, such as salaries and bonuses for the sales manager and support staff, may be allocated to each station/system/network based on its revenue as a percentage of total market/product revenue. Other companies want to see how much each individual entity contributes to net profit, and thus allocate only station/system/network-specific revenues to the individual entity. In this case, shared expenses are charged to an account that is not entity-specific. This gives companies the ability to easily separate station/system/network finances in the event there is a sale of some, but not all, of the pieces of a local cluster or business. There is no correct or incorrect approach to financial allocation, but consistently applying the *same methodology* is important in order to track and identify cost trends and analysis.

Typical Expenses

A discussion of typical expenses and their associated department classifications follows. Although companies may have different philosophies about the classification of some expenses, these expense classifications are typically standardized within an individual company.

Sales Expenses

These are expenses directly related to the generation of revenue for the operation. All things being equal, the cost of sales as a percentage of revenue should decline year-over-year because commissions and other expenses related to renewals are lower than those for new business. In reality, external events such as format changes and product additions keep this from happening. Costs typically included in sales are:

- **Added-value expense and merchandising.** These are expenses associated with a particular sales package or client advertising expenditures, and sometimes include trips and event tickets. Most companies cap this type of expense at 2 percent of the advertising revenue derived from any individual client.
- **Bonuses for management and account executives**, typically based on achieving predetermined revenue levels or revenue share of market goals.
- **Internet sales expense**, including third-party costs to outside sellers.
- **Meals** and **entertainment** for clients.
- **Fees** paid to national rep firms.
- **Outside-collection expense** for accounts written off to bad debt.
- **Ratings expense for third-party vendors** that track the size and demographic composition of audiences, including payments to research companies such as Arbitron, Nielsen, and Eastland. Some companies attribute this expense to the News/Programming Department because programming is directly related to audience ratings performance. As new tools of audience measurement emerge, one can expect this expense to continue to increase accordingly.
- **Recruiting** and **moving** expenses for sales staff.
- **Revenue-monitoring services.** These services track overall competitive market revenue, segmented by advertiser and by station.

- **Sales-commission expense**, including management overrides (this commission paid to sales managers for sales made by the local sales staff typically varies with sales revenues), and sales guarantees (a minimum guarantee of wages for a specific period, typically based on a fixed or sliding scale).
    - Sales commissions consist of two common types of policies: **payment on billing** or **payment on collections**. Both policies yield the same expense for income statement purposes; however, payment on collections creates a liability that is relieved over time as clients eventually pay their bills.
- **Sales-management salaries**, including director of sales and local and national sales managers.
- **Marketing materials** for the Sales Department.
- **Support staff for the Sales Department**, such as secretaries and administrative assistants.
- **Severance and termination costs** related to sales employees.
- **Software expenses**, such as demographic data, inventory and rate-management systems, and client database software.
- **Third-party commissions**, including Internet sales or other sales organizations.
- **Training expense.**
- **Travel reimbursement**, including mileage and out-of-town expenses.

News & Programming Expenses

The largest expenses in operating a radio or TV station typically involve news and programming. These are the costs required to produce or acquire programming, which includes the cost of labor. For example, in radio, a locally produced news/talk format typically requires many personnel, and hence has higher costs. Conversely, acquiring the rights to a nationally syndicated program can be very cost effective. The same is true for television stations, depending on the popularity of the program and available audiences. Sports programming typically requires rights fees, which can be extremely burdensome. Costs typically included in the News & Programming Department include:

- **Bonuses for on-air staff and other news and programming personnel.** Bonuses typically are based on achieving ratings or ranking goals for the targeted demographics of the station, and can equal or even exceed the base compensation.

- **Dues and subscriptions** for trade magazines and news publications.
- **Fees** paid for music, news, shows, or production material produced by a third party.
- **Music license fees** are a significant area of expense, and include payments to BMI (Broadcast Music, Inc.), ASCAP (American Society of Composers, Authors and Publishers), and SESAC. Recent negotiations of industry license agreements have made this a fixed cost (as opposed to a variable cost based on revenue achievement) for television and radio stations. More on music license fees is addressed in Chapter 14.
- **News** and **local traffic** services.
- **On-air giveaways.** Stations, in particular, attract listeners with contest prizes; these may include cash giveaways, merchandise, and/or trips. Some stations expense these items in the Marketing & Promotions Department.
- **Other outside services.** These can include programming consultant fees or voice talent by people other than station employees.
- **Production** and **recording** materials.
- **Programming research.** Research is performed periodically to keep a pulse on the popularity of the station as a whole, or on individual elements of the station's programming, such as the music or talent. The type and frequency of research studies can vary depending on the type of program content provided by a station.
- **Salaries for operations manager, program director, on-air and news talent, and production staff.** Other, more-format-specific positions include news writers and reporters for a news/talk situation; a sports station will employ sports reporters. Many of the news and programming employees are under contract with the station, and may be parties to noncompete agreements. Salaries for on-air talent often vary by format and day part, as well as market location. For example, in radio, a morning show personality probably will command a greater salary than that for evening or overnight talent, and a Top 40 station personality may command a greater salary than a personality at an easy listening station.
- **Severance and termination costs** related to news and programming employees.
- **Software fees** related to scheduling music or programming.
- **Sports rights fees.**
- **Syndicated programming.** After a rash of ownership consolidation in the 1990s, many companies tried to reduce their expenses associated with operations in

many of their smaller markets. Because talent costs are among a station's largest expenses, some companies have chosen to syndicate popular talent across many stations.

- **Talent fees for other station work, including remote broadcasts and appearances.** Employers are liable for payroll taxes associated with additional compensation employees receive when they are representing the company at events of this type. Policies should be in place to ensure that payments for these appearances are made through the Payroll Department, and not directly from advertisers. Talent-fee revenue billed to clients offsets talent-fee expense.
- **Training and travel expenses** associated with industry events.

Cable Network Programming Expenses

Typical expenses specific to a cable programming network include:

- **Amortization expense for internally developed programming** (these costs are normally capitalized and amortized over the life of the program).
- **Amortization expense for licensed shows aired** (these costs are normally capitalized and amortized over the life of the contract).
- **Consulting fees** to outside talent or research-and-development initiatives.
- **Launch support fees.** These are payments made by the cable programmer to the operator to help launch and promote the new channel on the local system. Launch support is not always offered; when it is, the support is amortized monthly over the length of the contract. As explained in Chapter 4, launch support is a contra revenue account. These fees are normally capitalized and amortized over the contractual term of the applicable distribution agreements as a reduction in subscriber-fees revenue. If the amortization expense exceeds the revenue recognized on a per-distribution basis, the excess amortization is included as a component of cost of service. This situation occurs when a network is attempting to increase the subscriber base, and justifies the higher cost of launch support by increasing revenues from advertising to pay for the growth.
- **Music license fees.**
- **Salaries** for the Programming Department.
- **Talent fees for other network work, including remote broadcasts and appearances.** Employers are liable for payroll taxes associated with additional compensation employees receive when they are representing the company at events

of this type. Policies should be in place to ensure that payments for these appearances are made through the Payroll Department, and not directly from advertisers. Talent-fee revenue billed to clients offsets talent-fee expense.

- **Training and travel expenses** associated with industry events.

Programming costs include in-house, original programming, as well as programming acquired from outside syndicators. The costs of original programming are capitalized (that is, they are treated as capital assets) and amortized over their estimated useful life. The various methods of amortizing these costs are: (a) the straight-line method, in which the same amount is written off for each period during the estimated useful life; (b) based on the number of expected showings; and (c) the accelerated method, in which proportionately larger amounts are written off in earlier periods of the asset's expected useful life (see discussion of "conservatism" in Chapter 3). The costs of acquired programming represent amounts paid or payable to program suppliers for the limited rights to broadcast programming. Exhibition rights under the licenses are generally limited to the contract period or a specific number of showings or runs.

As explained in Chapter 3 under Conservatism, original and acquired programming costs are stated at the lower of cost less accumulated amortization or estimated net realizable value. The programming inventory must be reviewed at year-end to ensure that the value on the books reflects the potential revenue that is going to be generated from each programming asset. If the potential revenue is lower than the value of the program in the books, then a write-down needs to be recorded.

Cable System/Operator Programming Expenses

Expenses specific to cable systems include:

- **Launch support fees**—As explained in Chapter 4, from the perspective of a cable operator, the accounting for launch support fees is the mirror image of what is done for networks (see above). Cash is paid by the programmer, either up front or as the launch is achieved, and is recognized as deferred revenue (subject to normal current-versus-noncurrent considerations). Such credits are amortized (usually straight-line amortization) over the term of the contract as a "contra," or reduction of programming expense.
- **Video-product expenses**—These are paid monthly to the network programming source, and are calculated on a rate per subscriber. These are determined based

on contractual arrangements, and are a function of a number of variables including type of programming offered (e.g., sports vs. religious programming), the size of the operator, length of the contract, and channel position. Additional terms may include volume discounts, limited basic carriage requirements, and system penetration and tiering incentives. (For more information, see Cable System-Specific Revenue Streams in Chapter 4.)

- **Retransmission consent**—Historically, over-the-air local television broadcasters were carried on cable lineups based on "must-carry" provisions established by the Federal Communications Commission (FCC). In 1994, the FCC gave stations a choice of being carried under the must-carry rules or under a new regulation requiring cable companies to obtain retransmission consent before carrying a broadcast signal. If a local broadcaster opts for retransmission consent, an agreement is negotiated between the station (or the station's affiliated network) and the cable provider. For many years, this has not been a cash transaction, but rather, an exchange of copyrighted program content for market distribution via cable. As of this writing, transactions increasingly involve cash payments.

Marketing & Promotional Expenses

Whereas Sales Department expenses focus on marketing to advertising clients, Marketing and Promotional Department expenses are those marketing costs that focus on capturing audience share.

- **Advertising expense**, including billboards, bus boards, TV spots, direct mail, radio, and Internet advertising. Because these expenses are typically very high, advertising expenses need to be evaluated carefully based on return on investment, which should include an analysis of how these costs impact ratings.
- **Event expenses** are those costs related to major events, concerts, and revenue-driven shows. Some companies include this expense in the Sales Department because it directly contributes to the revenue of the station/system/network. It is important to track these expenses by specific event, and to perform a financial analysis of an event's profitability to determine the likelihood of repeating it in the future.
- **Other promotional expenses** include those related to remote talent appearances, including the cost for banners, tents, and other items used for the appearance.
- **Premium items** such as T-shirts, banners, key chains, and other small items given away on-air and at station appearances.

- **Promotional software**, including prizewinner databases.
- **Salaries** for the marketing director, promotions personnel, and street team.
- **Station vehicle expenses**.

Technical & Engineering Expenses

The Technical Department is charged with keeping the station, cable network or cable system operating, and in particular, the station studios, production facilities, headends, and transmitters.

- **Contract payments for third-party** engineers, technicians, and consultants.
- **Circuit fees** for contract transmission to tower sites.
- **Power** and other **utility expenses** at transmitter sites.
- **Repairs** and **maintenance** expense for studio and transmission equipment that should not be capitalized. These can be large expense items.
- **Rents** at the transmitter sites.
- **Salaries** for engineers and technical staff.
- **Supplies** and **custodial property**.
- **Vehicle expenses** for engineers and technicians, including mileage or gas and maintenance on station engineering vehicles.

Information/Interactive Technology Expenses

As discussed earlier, emerging revenue areas present new areas of expense. If a station streams its signal via the Internet, there are significant rights fees to be paid for this additional "broadcast" of music and commercials. These expenses may fall into different departments, depending on the views of corporate and station management. However, all stations within a company should be *consistent* in reporting expenses in designated departments. Specific items included here are:

- **Computer hardware** and **supplies**.
- **Internet connection fees**.
- **Music copyright fees**.

- **Salaries** for webmaster or IT personnel.
- **Streaming fees.**
- **Talent fees.**
- **Web-hosting fees.**

General & Administrative Expenses

The General & Administrative Department includes all of the standard expenses of running the business end of a property; this category usually includes compensation to senior management and the finance functions. Other items typically included in this category are:

- **Bad debt expense** for uncollectible accounts receivable. Some companies expense bad debt in the Sales Department as a cost of sales. A predetermined percentage of net revenue is used as the calculation to allow for uncollectible accounts. This percentage used is based on the collection history of a particular station, and may vary from station to station within the same company and the same market. The reserve account should be reviewed periodically (at least annually) to determine its adequacy to cover potentially uncollectible accounts.
- **Banking fees**, including credit-card and scanning fees.
- **Bonuses** for senior management and finance personnel.
- **Charitable donations.**
- **Dues** and **subscriptions.**
- **Employer-paid benefits.**
- **Employer payroll taxes** associated with department payroll costs, including FICA, FUTA, and SUI.
- **FCC registration fees**, which are determined and billed annually by the Federal Communications Commission.
- **Franchise fees**, which are payments made by a local cable operator to the local franchising authority.
- **General Manager compensation** (some operations have both variable and fixed component to GM compensation).

- **Insurance** such as property/casualty, general liability, directors' and officers' liability (D&O), and broadcasters' liability.
- **Legal** and other **professional fees.**
- **Meals** and **entertainment.**
- **Office-equipment rental.**
- **Office supplies.**
- **Other taxes**, such as property, real estate, and local business taxes.
- **Postage** and **freight.**
- **Printing costs**, such as letterhead, business cards, and forms.
- **Salaries** for traffic, accounting, business office, and support staff.
- **Software license fees** for business systems, including general ledger, traffic, subscriber billing, workforce management, and accounts receivable.
- **Studio-** and **office-lease expense** should be amortized using the straight-line method for the life of the lease.
- **Telephone expenses**, including cell phones and LAN lines. Some of these expenses may be broken out and expensed separately by department. For example, telephone expenses directly associated with programming (request lines) or engineering (T1 lines, transmission lines, etc.) may be reported in those specific departments.
- **Third-party-provider expenses** such as payroll service fees.
- **Other miscellaneous expenses.**

Key Expense Controls and Measures

The first step in controlling expenses is breaking out budgets by line items. This allows the reviewer to flag areas that seem out of whack or inconsistent. Ongoing variance analyses during the year will also help flag areas in which expenses exceed budget.

For companies managing multiple stations, systems, or networks, leveraging the buying power of the group can reduce costs. For example, establishing a master contract with a national office-supply store that delivers the goods to the individual locations can reduce office-supply costs at the local level. In addition, employee paperwork for payroll and medical benefits can be administered electronically through new software packages. Television stations or groups may also look to group programming negotiations and the use of per-program options for music license fees (see Chapter 14) to control costs.

Centralizing operations such as sales, traffic, broadcast control rooms, and cable headends, as well as automating functions at multiple locations, may also reduce costs. Centralization minimizes both capital investment costs and operational costs. Outsourcing is often used to reduce the costs of personnel associated with the credit-and-collection functions, billing, and even in the Technical, Engineering, and IT Departments.

Local sales commissions should be evaluated on at least an annual basis. Compensation plans should be market competitive to avoid losing successful representatives to competing companies. Typically, different commission rates apply to (a) direct spot revenue, (b) agency spot revenue, and (c) nonspot revenue. Each commission plan should also be effective in achieving the goals of the particular entity, and should include measurable target goals and compensation incentives. For instance, rewarding sales in a particular revenue area can be executed only if that revenue can easily be identified within the traffic and billing system. Some companies have implemented rate-based compensation plans that focus on maximizing advertising inventory. Because turnaround time for submitting commissions to payroll is generally short, it is important that commission plans are easily understandable for both the sales and accounting staffs.

In Conclusion

This chapter outlines a number of the expenses typically incurred by electronic media companies. Although expense classifications may vary from company to company, they will generally be consistent among media properties of a single company. Accurate categorization of expenses is essential because it gives management the information it needs to evaluate the performance of each of its businesses. Knowing what's going out—and, more importantly, *why* it's going out—allows the savvy business professional to stay in control of an increasingly complex media operation.

6 Financial Systems for Broadcasting and Cable

Calvin Lyles, Jr., and Bruce Lazarus[1]

In another life, financial managers of modern media companies probably would be good jugglers because they have the ability to keep track of several things at the same time. To understand properly how a broadcast or cable operation stays in business, one needs a sense of "The Big Picture," and how various financial units, from sales and traffic to programming and operations, *interact* within a larger system. Master jugglers Bruce Lazarus, CEO of Cable Audit Associates (CAA), and Calvin Lyles, Radio Controller at Greater Media, Inc., show the reader the dynamics of some important financial systems.

Introduction

This chapter is designed to walk the reader through the various financial systems used by all business units throughout the broadcast and cable industries. It spans the systems that distribute the product or network all the way to the systems that provide stockholders with the financial results. The following departments within the organization at some point in time come into contact with the financial systems:

- Traffic
- Sales
- News & Programming
- Engineering
- Operations

---

1. The authors gratefully acknowledge Chris Bauschka, Industry Director of Communications, Oracle Corporation; Richard "Dick" Petty, SVP/Controller, Time Warner Cable; and Cyndee Everman, VP/Business Support Systems, Time Warner Cable, for their assistance with the cable billing, ERP, HRIS, and OSS portions of this chapter.

- Affiliate Relations (cable networks only)
- Customer Service (cable systems only)
- Finance

Traffic Systems

The traffic system impacts all of the various departments in radio and television, including cable television. Primary functions of the system are:

- Inventory management and commercial scheduling
- Invoicing and accounts receivable

Inventory Management and Commercial Scheduling

The traffic system manages commercial inventory for specific times throughout the broadcaster's programming. To the listener/viewer, placing a few commercial spots within a specified break time may seem rather easy—but multiply this procedure hundreds of times over for a typical broadcast day, and the challenge becomes daunting. The "trafficking" of commercials has improved in recent years thanks to enhanced inventory-management tools and better automation. As discussed in Chapter 4, the greatest source of revenue in radio and TV is spot advertising, so this system is extremely important in managing millions of dollars and thousands of commercial spots. In order to schedule a single commercial for an advertiser, key information is required on a traffic order. This information includes client, product, and estimate (called CPE by the advertising agencies that supply the information); the version of the copy or spot to air; product category; and commercial rate. To be recognized by most advertising agencies' spot buying and payment systems, the order *must* include the CPE information; if it is not included on all paperwork, the agency may delay payment until the discrepancy is resolved. A salesperson gathers this information in a proposal or order form. The order must include the flight (i.e., the dates that the advertiser wishes to run its commercials), and the specific day part requested. The sales order is then delivered to the Traffic Department, either by manually inputting the information or through means of an electronic interface, which is the preferred method for most national agencies. The system then generates a contract, which is printed and sent to the advertiser as confirmation of the order.

After the traffic person has scheduled all of the contracts, he or she reviews the entire inventory that has been sold for each individual date. The traffic person is responsible for ensuring that all spots ordered actually air as sold, and that there is proper separation between product categories and conflicting advertisers. Finalizing a traffic log is part art and part science. The person responsible must make sure that what goes out to the audience both works in the programming and maximizes station revenue. For instance, an advertiser who buys three spots within a one-hour program usually doesn't want them all in the same commercial break, and it's a bad idea to run a Toyota and a Volkswagen commercial back-to-back. The completed log is then delivered to the Programming Department for airing. The log includes everything for airing the program, including all commercial information as well as promotional spots.

In addition to the program logs, the traffic system provides data and reports for the sales-management team, finance team, and corporate management. Year-to-date reports on revenue pacing show how specific categories of revenue, sales executives' billing, specific product categories, specific clients, and so on are performing—month to month, and as compared to the same period in the prior year. The system also provides data for the average commercial unit rate (the mean price charged for a spot in a specific program or day part over a defined time period) and sellout history (a report that shows how a certain day part or program has done in terms of percentage of available spots booked). These reports can prompt management decisions by both providing early warnings of trouble and highlighting new areas in which to focus the sales team. In some traffic systems, pending orders and revenue can also be incorporated into the system. If this portion of the system is utilized, key personnel can combine business on the books information with information about pending business, enabling them to determine additional sales needed and percentage remaining to reach monthly sales goals.

Invoicing and Accounts Receivable

In radio and TV broadcasting, the traffic system also handles the billing of commercials and provides financial reporting of both the gross and the net revenue for the specified reporting month. The spots that were scheduled inside a specific date period are itemized on an invoice to display the client, product, estimate, rate, dates, and times that the commercial aired. The invoices are delivered to an advertiser or agency via mail or an electronic delivery service tied to the traffic system. They may also be delivered to a third party. (Electronic delivery is a work in progress for invoices that contain billing that is not traditional spot advertising.) A detailed billing report is generated to prepare a journal entry for entering revenue into the general ledger system. The billing reports are often also used for calculating monthly sales commission.

Emerging and nontraditional revenue sources—including web sites, concerts, downloadable content, event cost, talent fees, and many, many others—have created problems in the current traffic process because they do not utilize the same spot inventory as conventional commercials do. Many of these revenue sources are entered into the current traffic system as "nonspot" advertising so that the revenue and billing can be generated and included in the revenue and accounts receivable system. In the case of nonspot, the billing is displayed on an invoice as a single line item along with the revenue definition.

Once payments are received from the advertiser or agency, an accounts receivable clerk enters the payment into the system with the corresponding contract number, invoice number, or advertiser name to reduce the balance on the account. A summary report of all cash receipts for the month is generated from the system to prepare a journal entry for recording in the general ledger system.

Billing adjustments are sometimes required when a client short pays an invoice. Short payment may be the result of spots airing out of their intended day part, airing of the wrong spot, or an inaccurate sales-order entry. Bad debt adjustments are required when a delinquent client is turned over to a collections agency or the anticipated revenue is written off due to nonpayment. Billing adjustments and bad debt write-offs are entered into the accounts receivable system, and an adjustment report is generated in order to make the necessary general ledger adjustments to revenue.

The accounts receivable functionality of the traffic system retains all billing and payment history for an advertiser, agency, and salesperson. The collections person or accounts receivable manager generates an aging report to supply the Sales and Finance Departments with an accurate client history of payments and any remaining open invoices. The accounts receivable portion of the traffic system also enables the Collections Department to produce and send to the client statements that detail invoices and payments made. Because the system tracks all of the client's history, the credit manager can also utilize the system to review payment history, write-offs, and trends when determining whether or not to grant future credit to an advertiser. A summary accounts receivable report is generated at least monthly to tie back to the general ledger balance sheet statement.

Financial Systems for Cable Networks

As the name suggests, these systems are specific to the cable programming industry. Unlike the model in broadcasting, where most revenue is generated by advertising, cable programmers generally derive more of their revenues from cable operators, who pay for the privilege of distributing the programming.

The Network Affiliate Finance (NAF) Department within a cable program network is responsible for managing the billing and collecting of program license fees from cable systems. The fee is generally based upon the number of subscribers receiving the programming service. The subscriber fee can range from a few cents to several dollars per month. As explained in Chapter 4, more-complex license fee agreements may include terms based on specific carriage requirements, volume discounts, channel placement (channel number, or the "neighborhood" in which the channel is placed), penetration (percentage of potential customers who subscribe to the service), and packaging incentives. With billions of dollars collected and accounted for by the NAF group, the value of this group to the overall financial performance of the network is apparent.

The NAF staff's responsibilities are intertwined with the systems they use to perform these functions. The main responsibilities of the Network Affiliate Finance group include:

1. Launch authorization
2. System setup
3. Affiliate billing and cash application
4. Collection of affiliate receivables
5. Maintenance of affiliate database

Launch Authorization

The authorization process is usually initiated with the activation of a network-specific decoder at the headend receiving the network's satellite feed. This process usually requires the completion of a launch authorization form (LAF).

The LAF is completed by the cable operator or multiple-system operator (MSO) receiving the signal for the specified program network. A completed LAF will be faxed or emailed to the uplink/authorization facility. An approved LAF will initiate the authorization process to authorize a decoder to descramble and receive the network's satellite feed. The approved LAF will also be sent to the Network Affiliate Finance Department as the formal notification that a cable system will be launching the network service. In many cases, the NAF Department will be the final destination for the approved LAF. Much of the data contained in the LAF will be transferred to the affiliate database during the system-setup process.

The information contained on the LAF and required by the NAF group will include headend location, device type and number (for the piece of equipment authorized to decode the channel, sometimes called the "network decoder"), and technical contact information. In some cases, the LAF will include corporate contact and billing information. Having a fully completed LAF will ensure that the system is properly set up in the Network Affiliate Finance Department's affiliate database, assuring that future affiliate billing will go smoothly. It is equally important to maintain a thorough history of all LAFs. This is because, in many cases, cable operator/MSO payment backup (that is, the locations and subscriber counts sent with the monthly subscriber-fee payment) is not always consistent with the LAF. Reconciling LAFs with payment backup is probably the biggest challenge facing the NAF group.

System Setup

Following the authorization of a system, the NAF group will take all the information from the completed LAF and set up the system in the network's affiliate database. In addition to using the LAF data, the NAF will apply the appropriate cable operator/MSO contract terms, which include the monthly billing rates and applicable free periods. It is crucial to involve the Affiliate Sales and Legal Departments during the determination and the application of contract terms because these terms are often very complex.

Affiliate Billing and Cash Application

The Network Affiliate Finance group must have an affiliate database capable of executing both billing and cash application functions. The affiliate database must manage a large volume of data, duplicate and store similar transactions in the same manner, and provide an audit trail for each transaction. The more robust the affiliate database, the stronger the likelihood that the financial reporting will be accurate and verifiable.

The affiliate database can be developed internally, which has been undertaken by several large cable networks, or it can be licensed from a third-party software vendor. In either case, the development of this software never ceases—instead, its requirements continue to expand to adapt to a changing business environment.

Regardless of the software utilized, the major responsibilities of any NAF group will include:

- Processing MSO monthly subscriber updates
- Receiving and processing checks (lockbox)

- Generating and processing general ledger entries for funds received
- Reconciling cash received (lockbox batches) to cash applied (subscriber-management system, or SMS, batches)
- Matching systems billed to systems paid
- Generating payment-upload file
- Appling upload file (sub counts, rates, payments) to SMS
- Generating variance and exception reports
- Researching subscriber variances
- Researching rate and payment variances
- Applying changes/updates to SMS based on variance analysis
- Monitoring and verifying contract compliance
- Identifying and generating invoice adjustments
- Posting invoice adjustments to general ledger
- Reconciling accounts receivable

Neither a robust affiliate database nor a skilled staff is a substitute for what is commonly called "best practices." These are rules, processes, and procedures that govern how the Network Affiliate Finance group operates. Best practices apply to all users of the affiliate database, and especially pertain to individuals who are responsible for entering and maintaining data in the affiliate database. Consistency, accountability, and auditability are the foundation of best practices. Accurate financial statements will result from the implementation of best practices. For large networks that employ numerous staff to operate an NAF Department and maintain an affiliate database, without best practices, inconsistent management and operating procedures will degrade the network's operations because proper oversight will be extremely difficult and time-consuming.

Affiliate Database Maintenance

The Achilles' heel in any Network Affiliate Finance Department is affiliate database maintenance. Maintaining an affiliate database with current and accurate data is crucial to the department's success. Maintenance extends not only to the data itself, but also to the functionality of the affiliate database and the NAF Department's processes and procedures. Many cable networks, rather than staffing a department to maintain the affiliate database with industry data, prefer to purchase such data through third-party vendors that specialize in data collections. Although the accuracy of such data is questionable and may be out-of-date, it is cost effective and represents advancement for networks that choose not to staff internally to meet these challenges.

Cable System/MSO Customer Database and Billing

This system is specific to cable systems and cable operating companies. Again, unlike in broadcasting, where the majority of the company revenues are generated by the sale of advertising time, cable operators' primary source of revenue is generated from customer subscriptions to one or more levels of service. Known generically as cable billing systems, these are really key operational systems for cable operating companies. They house massive databases for all homes and businesses in a cable operator's service area, and track all customer transactions, beginning with the first contact between the company and the customer. These systems also provide data in a format that can be imported into the operator's general ledger system.

There are currently three main cable billing systems, along with a number of other systems, either homegrown or customized from other software packages. In the majority of cases, the billing system provides the first point of contact for data involving customer orders and changes. When a potential new customer contacts the cable company, the customer service representative initiates a service request in the cable billing system. Entering the service request allows the rep access to scheduling (the customer hears, "We have an opening tomorrow afternoon between 2:00 and 4:00"). Once the appointment is scheduled, the system sets up the work order and schedules the installation with dispatch, and provides the interface to complete installation—provisioning and the authorization of any so-called "addressable" cable services—resulting in the customer's being set up for monthly billing.

Most billing systems include individual customer credit limits. These give the cable company flexibility to offer optional services, such as video on demand (VOD), while maintaining appropriate internal controls against bad debt. Some cable systems have even set up their billing systems with real-time links to local credit bureaus, giving the company even greater control over credit decisions.

As one might expect, cable billing systems trigger the company's collections activities. The billing systems are also the first point of contact when a customer calls with a service problem. This makes sense because, in some cases, the service problem is a direct result of the customer's failure to pay for service.

Billing systems continue to change as cable company product offerings evolve and customers' habits change. Systems that 20+ years ago were capable of handling only basic and premium cable video services, with the occasional pay per view movie or event thrown in, now handle high speed data (Internet), telephone (the ability to provide usage-based billing varies among billing system packages), wireless, and commercial services. They can also include modules that interface with third-party fleet management or other management systems. Other options may provide web software front-ends that allow users to request service or report trouble online.

The variety of billing system options available depends upon the vendor. Some systems offer a service bureau approach—all data is managed through the provider's data centers. Other providers offer local servicing, which means the cable company houses the equipment and the data center. In some cases, the MSO may choose to set up its own regional service center to handle data from several cable systems. With some systems, the operator will have real-time access to customer and service data; in others, time-delayed data access. Statement presentation—the printing, stuffing, and mailing of customer bills—may be another option. And, as customers conduct more of their business online, billing companies are responding with electronic bill presentation options.

Each billing system includes financial reports along with the ability to automatically feed data to the company's general ledger. Reporting capabilities and data currency vary based on the billing company used and the options package(s) chosen. With all of the parts of a cable operator's business that they touch, cable billing systems are truly key financial systems for the operating companies.

Accounts Payable Systems

A company must have a sound method of selecting the proper vendor for purchasing products, equipment, and supplies; for approving vendor invoices; and for processing and recording invoices, expenses, and cash payments to the proper accounts in the general ledger system. A more detailed discussion of the types of expenses that relate to the broadcast and cable industries can be found in Chapter 5. All expenses and payments relating to entity operations are tracked through the use of accounts payable software systems. Accounts payable (AP) systems range from manual-type record keeping to very sophisticated automated systems, encompassing vendor history, purchase order approval, invoice matching, payment scheduling, batch entry controls, cash flow restrictions, security controls, and check signing controls. In the most basic AP system, an accounts payable clerk enters approved invoices into an accounts payable system. The information required for input is the vendor name, the invoice amount, and the account codes to which the payment is to be expensed. Checks are generated periodically based on the vendor's terms. Management regularly reviews the outstanding payables to determine future cash needs. The system generates reports that are imported into the general ledger system. The AP system is also used for expense analysis reporting, and generates reports that can be used for annual W-9 tax-reporting purposes.

Payroll Processing Systems

As one of largest expense components in broadcasting and cable, proper payroll processing is essential and leaves little room for error. A payroll system can range from manual periodic payrolls to large, sophisticated automation systems. The complexity of payroll processing is often underestimated. The Payroll Department is responsible for tracking information on hundreds of employees, both current and past. A typical payroll will include salaried employees, exempt and nonexempt, hourly employees, overtime, commissioned employees, talent fees, bonuses, and severance pay. Additionally, company benefits are usually administered and tracked through the payroll system, including vacation, sick time, insurance deductions, taxable fringe, 401(k) plans, and stock compensation. This process can be completed manually or, in most cases, through a sophisticated automated system or third-party service provider.

A basic automated-payroll software system retains pertinent employee information, such as address, Social Security number, tax preferences, and regular payroll deductions (including insurance and 401(k) deductions). Some systems are employee self-serve, which means that the employee can enter this information himself or herself. A payroll clerk receives time sheets and payroll information on a periodic basis, and enters the earnings and hours into the payroll system. The payroll system then calculates the pay and the required tax deductions, and generates the paychecks.

Most payroll software systems will automatically generate the reports required for the myriad of federal and state tax-reporting agencies. These include new-hire reporting, federal and state unemployment-tax reports, federal and state tax withholding, and W-2's. Some third-party providers also offer services such as check stuffing, direct deposits, tax filing, new-hire reporting, and W-2 and 1099 processing.

Additional reports generated from the payroll system provide analytical tools to management personnel. Reports such as salaries and hourly wages earned can be used to determine average pay per employee. Commission reports can be compared to revenue to evaluate compensation as a percentage of sales. A company may also want to track overtime and part-time hours per department.

All of the expense, payment, and accrual information obtained from the payroll ledgers must be recorded onto the general ledger in the appropriate department and expense lines. This can be done through manual entries, or through an automatic general ledger interface.

Fixed Asset Systems

Fixed assets typically include studio and camera equipment, towers, furniture, and other assets as defined in more detail in Chapter 7. Broadcast and cable companies require many fixed assets to operate. Fixed asset systems are used to track the assets related to the operation. As with the other systems identified in this chapter, the fixed asset system can either be managed by a simplistic manual or spreadsheet tracking system, or by use of more-sophisticated and automated-tracking software systems.

Assets are typically entered into the fixed asset system when entered into the accounts payable system at the time of purchase or project completion. The asset information included in the system is the value of cost of the asset, the life of the asset, the vendor, and the location. Assets can then be classified by various groups, such as asset type, asset location, general ledger account number, and personal property type. More-sophisticated fixed asset systems enable the company to keep track of model numbers, vendor contact info, and even maintenance records. The Finance Department is also responsible for updating the system when an asset has been sold or otherwise retired. The company can create periodic customized reports to assist in fixed asset inventory, insurance reporting, annual purchase and retirement activity, book value, and to handle reporting of real property and personal property taxes in certain jurisdictions.

Another primary use of fixed asset systems is to handle complex depreciation calculations. As discussed in more detail in Chapter 7, depreciation is the method of expensing an asset over time. Different depreciation methods are used for book and tax purposes. In some cases, a company may even need to calculate multiple depreciation methods for tax purposes alone. A sophisticated tool will apply tax rules simultaneously, and calculate depreciation for all of the various methods.

Book depreciation information obtained from the asset tracking system must be recorded onto the general ledger in the appropriate department and under the appropriate category. This process can be completed manually or through automation. The fixed asset reports are used to tie to the balance sheet accounts generated from the general ledger system.

General Ledger System

The Finance Department processes thousands of transactions, representing millions of dollars on a monthly basis. Every transaction will ultimately be reflected in the financial statements through the monthly accounting close. The Finance Department is responsible for providing the revenue accrual, monthly cash receipts, accounts receivable adjustments, revenue adjustments, accounts payable, and depreciation entries. All entries should be supportable by backup. Best practices, combined with a robust database and skilled staff, will assure that a proper audit trail accompanies each financial transaction. Because accurate financial reporting has become the cornerstone of Sarbanes-Oxley, the attention and resources devoted to the finance group reflects a business necessity to create an infrastructure to support its financial reporting responsibilities. For a more in-depth look at this topic, see Chapter 10.

The general ledger system is the central point of an operation's financial system and monthly close. The data and results obtained from the billing, accounts receivable, accounts payable, payroll, and fixed assets systems are all accumulated into the general ledger system to provide the financial results of the operations. The general ledger system provides key financial reporting to parent companies, investors, and analysts for activity during a specified period. The general ledger typically generates an income statement and balance sheet, as well as department cost detail and activity. The accuracy of the financial reports depends upon the accuracy of all the primary financial support systems outlined above that feed into the ledger.

Subsystems

As with any industry, various subsystems are sometimes utilized by companies to simplify, consolidate, and analyze various components of the business. The following are just a few of the systems that are commonly used by the broadcast and cable industries.

Pricing Software

Pricing software, often referred to as a yield management tool, is being employed in the industry to maximize advertising-sales dollars. The basic premise behind yield management is similar to pricing for the airline or hotel industries. There are a fixed number of commercial units available for sale, and the inventory is highly perishable. Once the time has lapsed, the inventory is no longer available for sale.

Through a complex set of calculations, pricing software relies on both current and historical information to determine how local commercial inventory demand is pacing against historical sellout levels. The software also reviews market revenue from prior periods based on pacing to prior years to estimate what the total revenue will be for the market. With all of this information, a pricing system assists management in establishing inventory pricing to maximize station revenue or to achieve goal.

A pricing system can be either static or dynamic. A static pricing model looks at all historical sales data, and determines a fixed, or "static," pricing grid—what the average yield should be based on current activity. In a dynamic pricing model, an actual proposal is first entered into the system. The system will look at each week and individual day parts ordered, and will set pricing for each order based on current demand or the ongoing "dynamics" of the marketplace, rather than on historical precedents. For example, the dynamic model will raise the rates when sellout levels reach various increments, such as 50, 60, or 70 percent sellout levels. The system reports the proposal as pending business, which further impacts the demand model and assists management in forecasting future revenue. Most available systems will integrate with station's traffic systems to pick up the most-current inventory and sales data.

Enterprise Reporting, Business Intelligence (BI), and Forecasting

Often referred to as enterprise reporting or business intelligence (BI), this software has gained importance due to ownership consolidation in the industry in the late 1990s, and the emergence of large companies with many stations/systems and multiple reporting units. Enterprise reporting software provides business intelligence and rapid information to identify trends and assist in forecasting and planning. For large organizations, these systems enable quick rollup of information. Based on OLAP (online analytical processing) or tree technology, it gives users the ability to drill down and analyze information from all areas of the enterprise, including human resources, general ledger, payroll, accounts payable ratings, market share, and other systems.

Human Resource Information Systems (HRIS)

Human resources represent the basic foundation of any organization. As larger companies have emerged through consolidation of the industry, tracking a company's employees has become extremely difficult, and yet critical to the success of the organization. Many of today's human resource systems can track and report on virtually thousand of employees. Also known as human resource management software (HRMS), such systems enable employers to input and retain much data on their workforce, including supervisors, training history, promotion and position history, performance reviews, skill and experience levels, technical certifications, and property assigned. These systems are typically integrated within a company's payroll system to reconcile and automate benefits within payroll and avoid duplicate entry. HRMS systems may also be included as part of an overall ERP package (see below).

The reporting capabilities of these systems are valuable. Some reports identify salary ranges by job type by consolidating employee information from various entities or divisions. The system can also track employee turnover ratios by division, station, or supervisor, and also assists in the management of benefits outside the payroll system—such as FMLA (Family and Medical Leave Act), benefit plans and coverage, COBRA (Consolidated Omnibus Budget Reconciliation Act), EEOC (Equal Employment Opportunity Commission), and OSHA (Occupational Safety and Health Administration).

Enterprise Resource Planning (ERP)

ERP systems encompass all back-office support functions in a single integrated package. Traditionally, ERP capabilities include accounting and finance, supply chain, and HRIS. Thus, an ERP package would include the finance elements mentioned earlier (accounts payable, fixed assets, etc.), as well as additional operational capabilities. The cable industry has used these enterprise-wide applications to help manage some of the unique aspects of a cable system.

Supply chain entails the management of physical equipment such as set top boxes, cable modems, and network equipment. The supply chain is the chain of events that brings equipment (supply) to the operator, manages the inventory, and distributes the inventory to its end location at a customer site or headend.

The first step in the supply chain is procurement. Inventory managers create *purchase orders,* which request inventory from the appropriate vendors. When the inventory arrives, the receipts are matched to the purchase order (P.O.) in the Accounts Payable Department, and the vendors are subsequently paid according to their contract. This is also known as the *procure-to-pay process.*

Once the inventory arrives, it is stocked in the warehouse. The stocking levels of each item in the warehouse are carefully monitored, and P.O.s are created when supply is low. The ERP application tracks the location and current quantity of each item in the warehouse as it is received, and will alert inventory managers when supply is low.

Finally, inventory is distributed to the appropriate personnel for installation. In the case of infrastructure equipment, it may have been ordered with a specific project in mind. In this case, the equipment is forwarded to the appropriate project team for installation. In the case of common, frequently used equipment such as set top boxes, the items will be stored and then pulled from inventory as needed. Field technicians will pull from this inventory to stock their truck each day as they complete work orders for customer installations and maintenance work.

Cable System Operational Support Systems (OSS)

Whereas ERP represents the back office of business operations, operational support systems represent those systems that interact directly with the cable network (i.e., the physical network responsible for delivering cable products to consumers). There are several key systems in this area.

First, provisioning systems are responsible for reserving the network capacity and activating a customer for a cable service, such as broadband or cable television. For example, the customer's cable modem will be registered with the network through provisioning, and likewise each set top box will be configured to show the appropriate packages and channels in each room of the house. Typically, an order will be passed from the customer care or billing system directly to provisioning in order to turn on the appropriate services.

Second, an engineering network (sometimes called "cable plant") inventory system is used to track the details of each device in the cable plant, including the overall network/plant design, device configuration, and device locations. These systems give network engineers as well as business systems a central point of record for all network information.

New Requirements in Cable

With the cable operators moving into new markets—such as digital phone, wireless, and commercial services—new requirements are being placed on the business and operational systems. From a front-office standpoint, agents now need to be able to place orders for multiple services at once, such as a "triple play" order of cable TV, broadband, and digital phone. This requires coordinating product definitions, pricing, promotions, and orders across these multiple service lines.

From a back-office standpoint, several of the new services, such as digital phone and wireless, require the gathering of usage data from the cable system network (or plant) itself for billing. For example, to offer a wireless service, an operator needs to know call details and minutes each month for each subscriber. Two systems work hand in hand to offer this capability. First, a mediation system ties directly to the cable system network to gather usage detail. Second, a rating engine receives the usage detail and calculates the charges based on usage. The rating engine then provides this detail to the core billing system, which invoices the subscriber along with the charges for other services.

In Conclusion

Broadcasting and cable companies have both common and unique financial systems. The need for advertising inventory management and commercial scheduling, for commercial invoicing, accounts receivable, accounts payable, payroll processing, fixed asset management, general ledger, and HRIS can be common, with minor variations, for radio, TV, cable programming, and cable operations companies. ERP systems may or may not be used in all industries, depending upon the corporate structure. Cable NAF and LAF systems are unique to cable programming companies. Both the specific cable OSS and cable SBS (subscriber billing systems) are, as the names suggest, unique to cable operating companies. As these industries continue to evolve and add new products/businesses, their financial systems needs will continue to change. What will not change, however, is the importance of systems that can quickly provide accurate financial data.

7 Capital Assets

Ronald Rizzuto, Ph.D., and Leslie Hartmann

Media companies have both big and small assets that contribute to the successful operation of the venture. Capital assets are the "biggies" that typically cost a lot, such as buildings and transmitters, and as a result, are regarded as a special financial category. Our authors look at the promises and pitfalls surrounding these expensive, long-term business investments. Dr. Ron Rizzuto is Professor of Finance, the Daniels College of Business, University of Denver. Leslie Hartmann, who also contributed Chapter 2, is Regional Director of Business Analysis for Entercom Communications.

Introduction

In previous chapters, you learned about expense items typical to a broadcast and cable operation. These items, as discussed in Chapter 3, are part of the operating expense section in the "Statement of Profit and Loss." We also touched on depreciation and amortization expense, as they relate to long-term assets that are displayed on a balance sheet, otherwise known as the "Statement of Financial Position." This chapter provides an in-depth focus on the long-term assets, specifically capital assets. It defines these assets, gives the reader specific examples of typical assets that fall within these categories, and addresses how these items are budgeted and acquired. In addition, it provides an overview of how these items are expensed or depleted through depreciation, retirement, and impairment. And finally, it addresses various measures of control that companies utilize from both an expense and security side.

Defining Long-term Assets

An asset has been defined as an item of economic value owned by the company or station. Current assets are those that are easily convertible to cash and are expected to be collected or consumed within a 12-month period; they include items such as cash, accounts receivable, and prepaid expenses or inventory. Long-term assets also have an economic value, but they are usually not easily converted to cash and have a life that exceeds a 12-month period. Capital assets, deferred assets, and intangibles are all components in the long-term-asset section of a balance sheet.

Capital Assets

Capital assets consist of tangible property—that is, property that can be seen or touched. Also sometimes referred to as "fixed assets," this category includes such items as property, equipment, and real estate. The broadcast and cable industries require large capital expenditures to maintain operations and remain competitive with up-to-date technology.

Most companies have policies regarding the dollar amount spent on an item in order to be classified as "capital." These lower limits may be as low as $500 and as high as $2,500, depending on the company's size and materiality thresholds. This is due primarily to the cost of tracking capital assets. For example, a calculator may well have a life in excess of five years, but it may cost as little as $20. The internal cost of capitalizing this item, determining and recording depreciation expense, and maintaining the item in inventory would simply not be cost effective. These smaller items are sometimes referred to as "custodial property," and are typically expensed in the period purchased.

There are major capital expenditures that relate to most industries. These include financial and ERP (enterprise resource planning) systems, discussed earlier in this book, and office facilities, such as a building or leasehold improvements.

The major items unique to radio and TV stations are purchases of transmission equipment, towers, studio equipment, satellite receivers, and remote broadcast vehicles. High definition technology for both radio and TV created the need to replace all analog transmission equipment with a digital delivery system, a very costly venture with an unidentifiable return.

The major capital expenditure items for a cable television operator are customer premise equipment (converters, cable modems, and telephone gateway equipment), headend and hub equipment, fiber-optic and coaxial cable, plant electronics (amplifiers and powering equipment), telephone switches, and advertising-insertion gear for multiple channels.

The major capital expenditures for a cable programmer are studio, cameras and editing gear, satellite equipment, and the video library.

Differentiating Capital Assets

Differentiating capital assets from repairs, custodial, or even current assets is often confusing. Some companies have had to correct their financial systems for unintentional misclassifications. As mentioned earlier, developing and maintaining a consistent policy is the best protection for avoiding such problems. Three examples are cited below:

- Repairs, which are expensed, are sometimes inadvertently treated as a capital improvement. Some examples of this would be a tube used in a transmitter or replacing cable lines. Although these items may meet the threshold for capitalizing, they may not actually increase the *long-term value* of the asset.
- Another area of distinction is when and how the items are purchased. For example, a chair may cost $250. On its own, it may be categorized as a custodial asset instead of a capital asset. But purchased as part of a set of ten chairs, the cost would be $2,500; hence the larger bulk purchase would be classified as a capital asset.
- A station may decide to give away a vehicle as part of an on-air promotion. Although the vehicle falls within the dollar amount of a capital asset, and may have a useful life of five years, if it is to be given away within the next 12 months, the economic life of the vehicle to the ongoing operation of the station is extremely limited. As a result, the vehicle would remain as a current asset, and be expensed over the duration of the promotion, or when given away.

Depreciating Capital Assets

In the case of a large capital project that spans several accounting periods before completion, the company will utilize a CIP (construction in progress) account until the project is finished and in service. Once an asset is placed in service, companies typically use a fixed asset software system to book the asset, and its "life" officially begins. The fixed asset software is used to track the assets, calculate depreciation, and populate the subsidiary ledger. Although many methods of depreciation are used for tax purposes (refer to Chapter 15), most companies utilize a straight-line depreciation method for bookkeeping and financial reporting purposes. The life (in years) is determined by the number of years that the asset is expected to have economic value. Most companies use a predetermined number of years, standardized for the various fixed asset categories, as identified above. These anticipated life spans are reviewed and updated based on technological changes. Once the life span for an asset is determined, the total cost of the asset, including installation and sales taxes, is divided evenly over the designated number of months, resulting in the monthly depreciation expense. The company will continue recording the depreciation expense until the value of the asset is fully depleted, or until retirement or impairment of the asset.

Asset Retirement or Impairment

When an item is replaced, an important decision is what to do with the replaced item. If the replaced item is of no practical use to the location, it may be abandoned, disposed of, or sold. These are examples of retirement, and each has differing tax consequences (see Chapter 15 for more on this topic). Some companies may choose to relocate or transfer the asset for use at another market. In any of these circumstances, the asset should be removed from the asset list to accurately reflect the company's current assets, and to avoid paying personal property taxes on the item. If the asset is sold, the sales proceeds—less any remaining value on the asset—are recorded as a gain or loss on the sale, depending on whether or not the sales price exceeds the current value. If the asset is retired or abandoned, any remaining value on the asset gets written off.

The asset list should be monitored periodically to identify obsolescence and assure that retired assets are both properly identified and properly depreciated on financial records. An impairment results when an item no longer has an economic value or its value has declined below the depreciated value. When impairment occurs, the asset must be written down and expensed on the books to reflect the reduced value; this may create a difference between the book and the tax value.

Expense Controls

One of the greatest challenges for the engineer is determining when to take advantage of new technology by purchasing new equipment. The engineer's primary responsibility is keeping the station, cable television system, or cable programming network up and running, so reliability is a major challenge. The decision-making process becomes a balancing act between becoming outdated with old but reliable technology versus jumping in early with new but untried technology that may prove unreliable. Moving too quickly to a new technology can be costly; sometimes new technology has not been well tested in the field, which can result in bugs or system failures. Technology advances quickly. Sometimes the decision to wait a little while results in significant price savings because prices typically come down when more competing product brands enter the market.

As mentioned in Chapter 5, large entities with multiple stations, systems, or programming networks leverage their size to reduce costs on capital expenditures. Often companies negotiate corporate discounts with particular vendors. Another expense-saving technique is to obtain even greater discounts through bulk purchases, such as buying many computers at one time. Controls for these capital expenditures should be similar to those applied to other purchases, such as vendor review and price quotes. Sometimes companies are more diligent in procuring capital assets because of the size, cost, and long-term outlook of these acquisitions. Additional controls include a capital purchase request form, which may supplement or replace a purchase order. Appropriate management approvals and review from higher levels of management are necessary.

One of the best expense controls is proper tracking and documentation of expenditures. Many companies document capital expenditures through a monthly or quarterly recap to track money spent to date and future purchase requirements. Some companies utilize bar codes for security and inventory purpose. Backup is maintained when items are capitalized to validate the capitalizable amount for the entire life of the asset, usually until retired.

Budgeting and Capital Planning

Capital budgets typically are done for each market or division. The engineer usually prepares studio and transmitter needs because of his or her expertise in these areas. Other department managers turn in annual request forms for their needs and general "wish list." If there is an IT manager, he or she will manage the division's computer hardware purchases. The division manager assesses all of the needs in the market and prioritizes first within the division. These budget requests are then consolidated company-wide, and provide a tool to review overall corporate needs and to set priorities based on factors such as urgency and ultimate return on investment. This reduces the need to constantly reassess and prioritize during the year. Because the number of homes to be passed by the cable plant has a direct impact on subscriber revenues, the cable system capital budget should be prepared before the cable system operating budget. Capital purchases for radio and TV stations may also impact revenue, albeit in a minor way.

Decisions about capital items may also impact the operating expense budgets for radio, TV, and cable. This is because items not approved for purchase may still be required for day-to-day use, and must be leased or rented. An example may be the requirement of a T1 line used in place of satellite receiving equipment. Other instances where capital expenditures may reduce operational expenses are in reduction of maintenance and repair costs on newer equipment, such as vehicles, transmitter, and studio equipment.

Different calculations are used to justify the purchase of capital items, as noted below. Typical ROI (return on investment) calculations include reduction of expenses or improved efficiency, such as improved sales productivity, enhancing sales opportunities, or the reduction in personnel.

Although capital assets are put on the balance sheet, and not immediately expensed to the operation's profit and loss statement, the purchase of capital assets requires the use of cash or borrowed money, and therefore the purchase can have a negative impact on an operation's cash flow. The ultimate control of the capital budget lies in the resources available to invest.

To evaluate such business proposals as marketing promotions, customer service improvements, training initiatives, new products/services, line extensions, acquisitions, plant upgrades, and the launching of new lines of business, it is imperative that one analyze the financial implications of these proposals.

There are three basic financial analytical frameworks for analyzing a business proposition:

1. Breakeven analysis
2. Payback analysis
3. Discounted cash flow analysis: net present value (NPV) and internal rate of return (IRR)

Each of these techniques will be discussed in more detail. Each method differs with respect to its level of analytical sophistication. Breakeven is the simplest of the methods, whereas discounted cash flow is the most sophisticated.

Anyone working in the electronic media industry should be familiar with all of these methods. For smaller, less-complex projects, a simple breakeven analysis may be sufficient. For more-complicated, less-expensive investments, payback analysis may be acceptable. However, for large long-term capital investments, an analyst will need to draw on the analytical horsepower of the discounted cash flow methods.

Breakeven Analysis

Breakeven analysis is an analytical method that focuses on the relationship between costs and profits at various levels of output. As the name implies, the technique is particularly focused on the output threshold required to generate zero profit or loss.

Knowing the breakeven profit level is useful to decision makers because it provides insight as to the likely economic feasibility of a project. That is, if the breakeven level of output is beyond the realm of possibility, then the decision maker knows that (a) the project is not feasible, or (b) the project will require a significant reconfiguration in costs/pricing in order for it to become feasible. In contrast, if the breakeven output is realistic, the decision maker knows the project/project design is reasonable, and the firm can proceed further with the project.

As the discussion above implies, breakeven analysis is not a technique that provides decision makers with a "go/no-go" decision rule. This technique simply identifies the actual hurdle that the project must surpass in order to be a viable investment. Once decision makers know this important benchmark value, they can use their expertise to determine if the breakeven value is attainable. It provides a "reasonableness test" for the decision maker, who then has to evaluate whether it is likely the company will achieve this level of response from the project.

Breakeven analysis can be applied to many different investment situations (e.g., promotions, new business opportunities, cost-saving situations, etc.). However, this technique is most effective under the following conditions:

1. There is one product.
2. There is one price for that product.
3. There is one variable cost (one variable cost driver).
4. All the variable costs are a function of output.

For example, assume a small cable company is evaluating whether to rent or buy a camera to meet its obligation to televise four local city council meetings annually. The company can rent the camera for $500 per meeting or buy a camera for $1,500. In this example, the camera is the "one product"; the "one purchase price" is $1,500; the "one variable cost" is the $500 rental fee that is driven by usage, and this fee is a function of output—that is, coverage of the city council meeting.

If one or more of these four conditions are not present, the investment decision maker will find that breakeven analysis does not have sufficient "analytical horsepower." In these situations, decision makers must upgrade their analytical methods to payback or discounted cash flow analysis.

Payback Analysis

Payback analysis focuses on the time frame required to recover the investment in the project. It is similar to breakeven analysis because it focuses on *recovery of one's investment*. However, payback differs from breakeven in the following ways:

1. Payback focuses on the length of time needed to recoup the investment, rather than on the output or revenue needed to recover up-front costs.
2. Payback is a projection- or forecast-based technique in that the decision maker has to make some assumptions regarding output or revenue in order to determine payback. Conversely, breakeven analysis solves for output or revenue.

3. Payback is flexible enough to handle multiproduct, multiprice, and multivariable cost-driver situations because it "dollarizes"[1] all of these factors. Breakeven, as noted above, cannot accommodate this level of complexity.

4. Payback analysis provides the decision maker with a "decision rule" (i.e., accept the project if the project payback is less than the corporate payback standard). Breakeven analysis provides a benchmark to focus on, rather than a decision rule.

A project's payback period is simply the length of time required to recover the initial investment. If a project generates equal annual cash flows,[2] the payback period is computed as follows:

$$\text{Simple Payback Method} = \frac{\textit{Investment}}{\textit{Annual Cash Flow}}$$

Annual Cash Flow

In order to make decisions with a payback analysis, decision makers must establish a "corporate standard" for investments. The standard is typically based on: (a) the experience of decision makers with respect to what has been successful in the past, and/or (b) the cutoff point after which, when projects are ranked from shortest to longest payback period, the capital available to fund investments is exhausted. Investments that have paybacks shorter than the "corporate standard" get funded as long as there is sufficient capital available. Those projects that have payback periods longer than the corporate standard get deferred.

Payback analysis provides more versatility than does breakeven analysis. However, it has limitations. These include:

1. The payback period ignores all benefits beyond that period.

2. As a consequence of ignoring benefits beyond the payback period, this analytical method biases capital spending in favor of projects that achieve short-term

---

1. Dollarize is used here to reflect a U.S. decision maker. Clearly, the monetary unit of value will vary by country.

2. Payback analysis uses cash flow rather than net income as a measure of benefits. Cash flow is defined as EBITDA (earnings before interest, taxes, depreciation, and amortization) before tax breakeven analysis and EBIT(1–t)+D+A (i.e., earnings after tax but before interest plus depreciation plus amortization).

profitability. Good long-term investments are difficult to justify when one uses payback criteria.

3. The "corporate standard" is a subjective measure. As a consequence, it is difficult to know when the standard needs to be updated because the experience on which it is based is no longer valid.

The limitations of payback analysis have made it necessary for decision makers to find more-sophisticated analytical tools such as the discounted cash flow methods: net present value (NPV) and internal rate of return (IRR).

Discounted Cash Flow Methods

Techniques such as NPV, IRR, and modified payback are analytical methods that incorporate directly the timing of cash inflows and outflows into project evaluation. By directly measuring the magnitude of timing of cash flows, these techniques are able to consider all the cash flows generated by the project, and they allow us to adjust the valuation of cash flows based on the time they occur. (The cash flows received earlier in the life of a project are weighted more heavily than those received later.) By directly adjusting for the time value of money, the discounted cash flow methods are able to eliminate the subjective short-term bias inherent in the payback method.

Assume, for example, that a cable operator is evaluating whether to build an extension to provide service to a hotel. In this case, the decision maker knows that the hotel will pay a fixed monthly rate per guest room over the term of a three-year contract. A discounted cash flow analysis will give the earlier payments more weight than those in the final months of the agreement.

NPV and IRR are preferred methods over payback because:

1. They directly consider the value due to the timing of the receipt of cash flows.
2. They eliminate the bias in capital spending against long-term projects.
3. They consider all the costs and benefits of a project.
4. They provide decision rules that are founded in principles of economics rather than on subjective judgment.[3]

---

3. NPV and IRR decision rules are essentially the same decision rules used in microeconomics. Namely, invest if marginal revenue is greater than marginal cost.

Time Value of Money (TVM)—Background

The NPV and IRR methods are founded on the concept of the time value of money (TVM). Intuitively, this concept is simply the idea that individuals prefer money "sooner rather than later." Funds received today are more valuable than the same amount of money received in the future. This is because individuals can invest the money today and have the original sum *plus interest* in the future.

Think of a certificate of deposit (CD) with a 5 percent simple rate of return for a 12-month investment. If you deposit $1,000 today, you will have $1,050 a year from now.

Present Value Analysis

In many TVM problems, rather than solve for the future value, a manager may be interested in the amount one has to invest today in order to achieve a future financial objective. These backward-looking TVM problems are commonly labeled "present value" problems.

The process of determining present value is simply the *reciprocal* of the future-value problem.

Net Present Value (NPV)

Financial analysts have historically had a preference for analyzing projects on a present value basis. That is, rather than compare which projects provide the highest future value; they prefer to compare the present value of alternative investments. One key reason for this is because it is quicker to calculate present value.[4]

The net present value (NPV) method discounts the future cash flows at a discount rate,[5] and subtracts the cost of the project. Because the investment is already in present value terms, the process of discounting (present valuing) the cash flows allows a comparison of costs and benefits in today's dollars.

---

4. In order to compare future values, both the initial investment and the project's cash flows need to be future valued. With present value analysis, only the project's cash flows have to be discounted, because the investment is already in present value terms.

5. The discount rate, k, is the company's cost of capital. This is a weighted average cost of the firm's debt and equity securities used to finance capital projects.

Think of the CD example above. In that case, \$1,050 a year from now has a net present value of \$1,000—that is, it is worth \$1,000 in today's dollars.

Internal Rate of Return (IRR)

Whereas the NPV method summarizes an investment from the perspective of the increase in wealth that a project provides for the organization, the IRR method focuses on the percentage annual rate of return that the project generates. IRR and NPV are analytical methods that lead to the same accept/reject decisions. The fundamental difference between the methods is that NPV uses a profit metric (i.e., NPV gives "dollars increase in wealth"), whereas IRR provides an annual percentage return on the capital invested.

Using Financial Measures in Making Capital Investment Decisions

The financial analytical techniques discussed above help decision makers understand the financial viability of a project. Clearly, the likely profitability of a capital project is a key factor; however, it is not the only criterion used in making capital investment decisions. Sometimes projects that are *not profitable* still get funded because of other critical considerations, such as:

1. It is needed to maintain the infrastructure of the business (e.g., the building's roof needs to be replaced).
2. It is needed to meet a competitive threat (e.g., a competitor is expanding its channel capacity so as to offer more high definition television services, hence we need to make an investment for defensive reasons).
3. It is needed to test the viability of a new idea (e.g., in order to test the market viability of on-demand advertising, the company may want to experiment with this investment in one market before considering it for the entire corporation).

In Conclusion

Capital investment decisions are *business* decisions as well as simple finance decisions. Hence, these other factors need to be considered seriously. The financial metrics noted in this chapter, however, provide the framework for decision makers to determine if and when they need to make trade-offs in evaluating capital projects.

8 Key Performance Metrics

Samuel D. Bush

No sane person would use a stethoscope to see if someone had a fever—because it's the wrong device. Just as doctors use specific diagnostic tools to evaluate someone's overall health, so do financial managers have their own set of measuring tools, called key performance metrics, to calibrate the "heartbeat" and other vital signs of a business. Sam Bush, Senior Vice President, Treasurer, and CFO of Saga Communications, Inc., provides a highly detailed look at the many ways the financial performance of a broadcast or cable company can be diagnosed.

Introduction

Key performance metrics are valuable tools for analyzing one's own company with the goal to improve performance, or for appraising future growth of a competing company to determine the overall value of a possible merger or acquisition. Key performance metrics for the media industry cover a lot of territory, including assessments related to a company's income statement, cash flow statement, balance sheet, operational market data, and several other reports that overlap in different areas. Most of the metrics regularly reviewed are based on financial statements, and depend on a company's own financial records because competitors' financial statements are seldom available. This restriction is not always absolute, in that public media companies are required to periodically release overall performance numbers. Although not as good as market-specific performance numbers, these public disclosures often can provide the ability to do high level comparisons among companies in the same industry.

Tools and Resources

Several outside resources provide industry- and market-performance information. These include the Broadcast Cable Financial Management Association (BCFM); BIA Financial Network (BIAfn); SNL Kagan; Miller, Kaplan, Arse & Co., LLP; the National Association of Broadcasters (NAB); the various state broadcasters associations; the

National Cable & Telecommunications Association (NCTA); the Cable and Telecommunications Human Resources Association (CTHRA); Veronis Suhler Stevenson (VSS); various Wall Street investment banking firms; and many industry-specific publications. Data available include ownership and contact information, pending and completed transactions, estimated revenues, growth statistics and trends, market demographic information, technical data, benchmarks, projections, and historic salary information, along with a supply of other information helpful in evaluating and comparing media properties. Links for most of these (and other) sources can be found on the BCFM web site (www.bcfm.com). A quick look at the individual web sites will give the reader an understanding of the specific information available.

Key Performance Metrics—General

Cash Flow Margins

The various forms of cash flow margins are designed to measure the business's operating performance by expressing income as a *percentage of revenue.* This can serve as a great indicator of change over time, and allow the analyst to focus on the reasons for this change. Once the reasons have been identified, a manager can make an informed decision about correcting a problem or capitalizing on an opportunity. Two forms of cash flow measurement are EBITDA (earnings before interest, taxes, depreciation, and amortization) and free cash flow (FCF). Broadcast Cash Flow (BCF) is covered later in this chapter. This margin is relevant only to radio and television operations.

EBITDA is calculated as Net Income + Interest Expense + Tax Expense + Depreciation and Amortization. EBITDA is useful in making comparisons about overall company performance within the industry as well as outside the industry.

Free cash flow (FCF) generally approximates cash generated or expected to be generated over and above operating and financing needs. FCF takes into consideration both the income statement and balance sheet in comparing how much actual cash a company has available for potential acquisitions, to pay dividends (or other forms of equity compensation), to make nonscheduled payments on outstanding debt, to buy back stock from existing shareholders (generally applies only to public companies), and for return of capital to owners of privately held companies. The calculation of FCF is Net Income + Depreciation and Amortization + Deferred Taxes (taxes owed but not paid in the current period) +/− Other Income or Expense + Noncash Compensation − Capital Expenditures. In other words, FCF is simply EBITDA *minus* capital expenditures, *minus* cash interest (interest payments made "in cash" during the period), *minus* debt-service payments (payments of borrowed principal) *plus or minus* other income/expense.

Liquidity Ratios

Liquidity ratios provide a measure of a company's capacity to handle its short-term obligations as they mature. The most common is the current ratio, which is calculated by dividing current assets by current liabilities. A current ratio *equal to 1* means that a company's current assets are exactly equal to its current liabilities; this tells management that if the company is to continue to meet its short-term obligations in a timely manner, there is no margin of error in the current asset accounts. The greater the margin *above 1,* the better a company's "cushion" to protect itself from the unexpected. Take, for instance, the case in which a client experiences financial difficulties and begins to slow down payment on invoices. This results in a deterioration of the company's accounts receivable collection—and ultimately, this means less cash to pay current liabilities on time. A current ratio *below 1* is a warning sign that the company is experiencing current asset or current liability problems that could impair its ongoing ability to operate.

Another liquidity ratio is days sales outstanding (DSO). The DSO ratio is calculated by dividing a company's total accounts receivable by the average net sales *per day.* This gives management the average number of days of sales remaining unpaid from advertisers. For instance, a company that on average sells $10,000 worth of advertising per day, and has total accounts receivable outstanding of $642,000, has a DSO of 64.2 days. This means that on average, it takes 64.2 days for the company to collect its receivables. (Some firms figure DSO by eliminating cash-in-advance sales from the calculation in order to measure the overall effectiveness of the Collections Department.) What is an acceptable DSO? This depends on a number of factors, including the segment of the industry one is in, the markets in which a station or system operates, and the company's credit policies, as well as a host of other factors. To use this liquidity ratio effectively, first determine the *norm* for the station, market, company, or segment of the industry. Once this benchmark is established, management can use the DSO ratio to evaluate whether station, market, or company performance is above the norm and thus puts it at risk for a higher level of bad debt if appropriate actions are not taken. BCFM provides participating members with periodic DSO information. For more information contact the association.

Return on Assets

Business analysts often want to know how well a company takes advantage of its assets. Return on assets (ROA) is calculated by dividing net income by total assets (or, more strictly, net income plus interest expense – net of income tax savings – by total assets). Investors often look at this measurement when evaluating the companies in which they are considering making investments. It is only common sense that an investor wants to see his or her money used in the most cost effective and profitable manner. Internally, management can use this same concept to help determine where within the organization to invest funds to maximize return on the investment. For instance, a company may have divisions that operate in some or all of the following areas:

- Radio stations
- Television stations
- Cable systems
- Television networks
- Cable programming
- Billboards

On the other hand, a company may operate in only one of these areas, but in markets that vary dramatically in size. By using an ROA calculation for each "operating" segment, management can determine where available funds should be invested in order to obtain the maximum return.

A similar but slightly different measurement that reveals how well a company uses its assets is return on investment (ROI). The concept is the same as ROA, but is calculated by dividing the net return or expected net return (gain from the investment minus the cost of the investment) by the cost of the investment. In this case, the company can tell which investments are worthwhile and deserving of continued support, and conversely, which investments are disappointing in their productivity.

Segment-Specific Metrics

In addition to the key performance metrics described above, there are a number of measurements specific to either broadcasting or cable.

Segment-Specific Metrics—Broadcast

Power Ratios

Although used more often in radio, this ratio can also be applied to television and cable as a means of determining how efficiently a station or cable network converts ratings into revenue. More specifically, this ratio measures the company's share of audience compared to its share of advertising revenue spent in the market. The ratio can be calculated only when ratings and revenue data relative to market competitors are available, and is most often used in radio to evaluate the relationship between a station's share of revenue and its share of the 12-and-over population of radio listeners in that market. Audience information is reported by a ratings service, such as Arbitron. Power ratios can also be calculated using different demographic groups, such as adults ages 25 to 54.

The power ratio is calculated in two steps. The first step is to obtain the total revenue in a market and multiply it by the specific audience share (either 12+ or 25–54) for that station, which yields a revenue amount based on audience share ("audience share revenue"). Then divide the actual revenue for the station by the "audience share revenue" calculated in step one. The result is a number either greater or less than 1. If the station's actual revenue is less than the "audience share revenue," then the power ratio is less than 1. If the station's actual revenue is greater than the "audience share revenue," then the power ratio is greater than 1.

Take, for example, a situation in which total market revenue is $10 million, and a specific station has a 12+ audience share of 15 percent. On a 1:1 basis, the station should expect $1.5 million in revenue ($10 million × 15 percent). Now assume that the station's actual annual revenue is $1.2 million. In this example, the station is actually converting less than 1:1, and thus is not converting revenues based on all of the audience ($1.2 million divided ÷ $1.5 million = 80 percent, or a power ratio of 0.8) against total market revenue. This would indicate that another station in the market probably has a power ratio greater than 1.0 to make up the difference.

A station's power ratio can be influenced by a number of factors, including the station's programming format. In radio, it is generally accepted that a news/talk station will have a higher power ratio than a CHR (contemporary hits radio) station because advertisers generally believe that reaching the listeners of a news/talk station is a better consumer target than CHR listeners. The news/talk audience is perceived to be older, more affluent, and in control of more disposable income. In general, the higher the power ratio, the better a station's sales staff is performing. Although a higher power ratio is good, this could present some problems in the future. As other stations improve their sales performance, the station with the initial high power ratio will likely begin to perform more in line with its audience share. In this case, its revenue growth could be lower than that of the market in general, and may even result in revenue declines as the station's power ratio declines.

Broadcast Cash Flow Margin

Historically, the most common cash flow margin used in the broadcasting industry was referred to as broadcast cash flow (BCF). Broadcast cash flow is simply revenue minus operating expenses. (BCF is sometimes referred to as station operating income.) The BCF margin is very simple, and is calculated by dividing an individual station's or station group's operating income by the same station's or group's gross or net revenue. This margin focuses totally on the operating performance at a station level, and ignores any corporate overhead, interest, amortization, depreciation, and taxes. It can be used to compare individual station performance against overall industry norms, or against specific station competitors. In addition, a company that has radio (or television) operations in several markets might use this ratio as one way to determine how well the management team in one particular market is performing versus its peers in other markets. Any analysis like this, however, must also take into consideration the differences between markets, including economic base (service, industrial, military, educational, etc.), population size, geography, and ethnicity (among others).

Table 8.1 shows BCF margins for a hypothetical company's stations in three different markets.

	MARKET A		MARKET B		MARKET C	
Net Revenue	$2,960,000		$2,300,000		$1,100,000	
		% of Net Revenue		% of Net Revenue		% of Net Revenue
Technical	123,000	4.2%	75,000	3.3%	15,000	1.4%
Programming	580,000	19.6	243,000	10.6	205,000	18.6
Sales	805,000	27.2	615,000	26.7	185,000	16.8
Advertising & Promotion	167,000	5.6	62,000	2.7	45,000	4.1
General & Adm.	800,000	27.0	580,000	25.2	325,000	29.5
Total Expenses	$2,475,000		$1,565,000		$775,000	
Broadcast Cash Flow	$485,000		$735,000		$325,000	
BCF Margin	16.0%		32.0%		30.0%	

TABLE 8.1 *Comparison of Broadcast Cash Flow Margins*

What are some of the things we can learn from a BCF comparison? First, one can compare the relative percentage of net revenue a station is spending in each major expense category in the various markets. For instance, technical expense is only 1.4 percent of net revenue in Market C, whereas it is more than double that in Market A (4.2 percent) and Market B (3.3 percent). What could this mean? It could be as simple as a situation in which the company operates only one station in Market C, whereas it operates two or more in each of the other markets. It could mean that in an attempt to increase BCF, the manager in Market C is putting off routine maintenance and creating a situation that may cost the company more in the long run. One other quick example: The programming in Market B is 10.6 percent of net revenue. At the same time, Markets A and C are relatively similar to each other, at 19.6 and 18.6 percent, respectively. This would certainly be a situation worth investigating. Larger markets might require more investments in programming and promotion. Typically, markets of similar size will have comparable ratios.

Segment-Specific Metrics—Cable

Cable system analysis has become increasingly complicated as cable operators have expanded services beyond providing traditional analog video channels. Cable companies now can offer customers a combination of on-demand services, data services (Internet), telephone, wireless, and commercial packages—and they continue to explore new ways to use their relationship with their subscribers. Let's first look at some general metrics used for the entire system operations.

Homes Passed—Is a measure of the number of unique residences (homes, apartments, etc.) within the scope of a cable system's franchise that are technically "passed" by cable wires, and thus have the *opportunity* to subscribe to some or all of the system's products. More specifically, "passed" means that the cable system actually has cable running either underground or aboveground adjacent to the residence, and thus the residence can easily be connected to the cable system's headend once a subscription is obtained. Homes passed is an interesting statistic because it reveals a number of things about a cable system's future opportunities. For instance, if the total number of unique residences in a cable system's franchise area is 75,000, and the cable system has "passed" only 40,000 of them, one would expect that the system could grow its potential customer base by installing more infrastructure, thus giving the additional "homes passed" the opportunity to subscribe. This may not be as easy as it sounds. For one thing, the system may be in a rural area that has only a small nucleus of population (for instance, in the town center), with the remaining franchise population spread out over hundreds of square miles of countryside. It is easy to understand that if a cable system can "pass" 100 unique residences for every mile of cable it installs, the potential profitability is significantly greater than in the more rural area, where the system "passes" only 5 (or fewer) unique residences per mile.

Annual OCF or OFCF per Home Passed—Allows a comparison of the operating cash flow (OCF, similar to BCF in broadcasting) or operating free cash flow (OFCF—actual cash available after capital expenditures, taxes, interest, corporate overhead, and all other expenses are deducted) per the number of homes passed. Basically, this is a measure of a cable operator's efficiency in using its capital to build out a single system or multiple systems. OCF is defined as the total revenue minus the total direct costs and total operating expenses. OFCF, on the other hand, includes the additional subtraction of the capital expenditures from total revenue.

Bundle Discount Percentage of Monthly Recurring Revenue—Allows the system to evaluate how effective its *packaging* of all of its services is in generating additional revenues. Most cable operators discount individual services (video, data, and telephone) when they are bought as part of a bundled package. This can be an important metric because a system prices its services on an individual basis as well as on a packaged basis in order to maximize profitability.

Now let's look at a few of the metrics used to measure performance of the traditional video-service product. These measurements can also be applied to the data- and telephone service products by simply changing the measurement definitions. This will be explained further after the video metrics are described in some detail.

Basic and/or Digital Net Gain—Measures the net new subscribers that a cable system has acquired for either its *basic cable service* or for its *digital cable service*. This calculation simply takes the number of subscribers that the cable system has for either basic or digital service at the *end* of a given time period (usually monthly, quarterly, or annually), and subtracts the number of subscribers for the same service that the system had at the *beginning* of the time period. If a cable system is growing its subscriber base, this calculation should result in a positive number, but it can vary widely by the system's location (for example, areas with transient populations, including students or so-called snowbirds, can cause cable systems to lose subscribers at specific times of the year), as well as by how long the system has been in operation. If the calculation is negative, the cable system has lost more subscribers than it has signed up during the measurement period. This could be one sign of customer service-related issues that need to be addressed. The system operator may also want to look carefully at the specific gains (or losses) of the basic customers versus the digital customers. This can help management determine the success of specific marketing efforts aimed at increasing the number of digital customers versus the number of basic customers. Historically, digital subscribers generate substantially more revenue than do basic subscribers.

Basic and/or Digital Churn Rate—Measures the percentage of customers who disconnect during a specific time period. It is calculated by dividing the number of customers who disconnected video services during a specific time period by the number of customers who subscribed to the service at the beginning of the time period (usually a month). This is a very important measurement because of the significant costs in acquiring and setting up a new customer. These costs include marketing (advertising and special promotions aimed at getting potential customers to sign up for service), customer service (actual process of signing up the customer, including getting the relevant information as to where the service is to be provided and how it is going to be paid for), and the cost of sending a technician with all the necessary equipment out to the service site to install the new service. A reduction in churn rate can greatly enhance a system's profitability.

Analog and/or Digital CMPU (contribution margin per unit)—Is calculated by adding either the complete basic service or digital service revenue to the equipment rental revenue and then subtracting the service costs—the fees the system incurs to obtain programming service from networks (e.g., the fee per subscriber that a system pays to carry ESPN or the Discovery Channel)—and dividing the result by the total number of subscribers to the level of service being measured. This measurement is exactly what it says. It is the contribution margin per unit, in other words, the *amount of financial contribution that each unit of service purchased by a subscriber makes toward the system's total revenue.* Additionally, the system would look at the breakout of "tiers" of services within both the analog and digital subscriber categories. For instance, within the digital tier, one might find a "basic" digital subscriber who may or may not subscribe to the HD (high definition) digital service or the DVR (digital video recorder) service. Just as digital subscribers generally mean more revenue to the system than basic analog subscribers do, both HD subscribers and DVR subscribers provide substantially higher revenue opportunities than do the "basic" digital subscribers. Subscriptions to these higher levels of service generate more revenue for the system, and they usually require the rental or purchase of additional equipment, again providing for additional revenues to the system.

Applying the Metrics to Data and Telephone

To apply the net gains or churn metrics to telephone or to other services, simply change the data being measured from basic or digital subscribers to VOD (video on demand) subscribers, HSI (high speed Internet) subscribers, telephone subscribers, and so on. Measurements in the data-service area might also break down the net gains into tiers based on the type of service being provided. Data packages may be priced based on the speed of the connection—the faster the speed, the greater the cost to the subscriber. Higher subscription fees generally mean better profitability to the cable operator.

When calculating some or all of these metrics related to the cable industry, it is important to remember that there is some overlap among these measurements. For instance, when calculating total subscribers to the video services of a cable system, be sure to not *double count* the number of customers. All digital subscribers are generally required to be basic subscribers as well—thus, a complete count is not as simple as adding the total number of basic subscribers to the total number of digital subscribers. This issue is also relevant to the consolidated count of a cable operator's customers. For example, some customers may subscribe to all three (video, data, and telephone) services, whereas others may subscribe to just one or two of the services. Sometimes these separate services purchased by a single subscriber are called revenue generating units (RGUs).

In Conclusion

Always keep in mind that metrics are only as good as the manner in which they are used. Many a good operator, analyst, or investor has taken valid data and *applied it incorrectly*. There are many management questions to which data alone cannot provide the answers. Raw numbers are valuable only when combined with the appropriate analysis and interpretation. One must utilize the multitude of valuable resources that are available within the broadcast and cable universe to get a better understanding of the industry or industry segment that is under review. Combining good data with good analysis and interpretation will provide the necessary information to make good operating and investing decisions.

9 Cash Flow Measures and Reports

Anthony Vasconcellos

On a personal level, bouncing a check can be an embarrassing experience for anyone. Now imagine that you are the head of a major business, and suddenly you are receiving irate phone calls from vendors complaining that your checks will not clear. What's the problem?—cash flow! You may have money "on the books," but these numbers do not reflect a financial reality in which you can pay the bills. Anthony Vasconcellos is Executive Vice President and Chief Operating Officer for Regent Communications, and he explains concisely some of the factors that influence the all-important cash flow of a company.

Introduction

Maximizing cash flow is the ultimate goal of all businesses, because cash is the lifeblood of any enterprise. Reported increases in profit, market share, audience share, or revenues are meaningless if a business is unable to pay salaries and dividends and pay off debts with "real money." In recent years, managers, bankers, and potential investors have placed more emphasis on cash flow as an overall measure of a company's true *value*. Taken from the organizational to the personal level, think in terms of your own salary. You can work two weeks and "earn" $2,000 (accrual accounting), but it doesn't do you any good until you actually receive the paycheck and funds can be drawn from this sum. In simple terms, if a business doesn't have the cash in the bank to pay vendors or employees or is bouncing checks, that business has a cash flow problem.

Relevance of Cash Flow

Although a very simple concept to understand, cash flow is, in fact, difficult to manage. In the past, *working capital*, rather than cash flow, was the focus of attention. However, because of complexities associated with accrual accounting and the application of more sophisticated accounting principles, a more concrete, simplified cash flow method of analysis began to gain acceptance as a true measure of performance. Today, the Statement of Cash Flow is one of the most vital financial reports an organization provides to its management, lenders, and investors. This intense scrutiny of cash flow stems from one crucial fact: *as cash flow increases, so does the value of the business.* As evidence of the importance of cash flow, sales prices of broadcast and cable properties are typically based as a multiple of cash flow (also called operating income—this reflects earnings before interest, taxes, depreciation, and amortization). For example, investors may determine that the market value of a station for sale in a certain part of the country is *10 times cash flow* (before corporate overhead costs).

What Is Cash Flow?

Cash flow is nothing more complicated than calculating the difference between the bank balance at the beginning and the end of a specified time period. The key to understanding cash flow is, of course, the thorough analysis of the components contributing to this difference.

- Net income (with noncash expenses added back)
- Accounts receivable
- Prepaid expenses
- Taxes
- Fixed asset purchases
- Financing

Notwithstanding a sudden—and welcome—increase in sales, cash maximization is best achieved by attention to these elements.

For additional information about cash flow statements, see Chapter 3.

Enhancing Cash Flow

Net Income

In cash basis accounting, the *net income* of a company becomes its cash flow. As a result, maximizing cash is achieved by increasing revenues and minimizing expenses. In accrual accounting, sales generate the receivables, which need to be collected, and expenses generate the payables, which need to be paid. Net income then becomes the starting point when collecting cash from customers (i.e., advertisers and subscribers). In other words, even after having had an exceptional sales month, a business may be short of cash. If, for example, receivables balloon or cash goes to pay down balances with suppliers, cash balances may decrease even after a month of great sales.

For simplicity's sake, the discussion in this section is limited to maximizing cash flow from key components of the overall analysis.

Accounts Receivable/Cash Collections

Accounts receivable, that most basic and important of assets, is not given the attention it deserves by some organizations. In addition to the obvious cost of financing receivables, the risk of late-paying customers turning into collection or legal problems adds special significance to receivables.

A quick example shows the cost of financing customer accounts. If current receivables indicate days sales outstanding (DSO) of 90 days, and monthly net revenue averages $2,000,000, a company is financing customers to the tune of $6,000,000 (90 days of revenue in receivables). If collections were improved by 20 days, the average receivable balance would decrease by approximately $1.3 million. Using an interest rate of 10 percent, the 20-day improvement would result in a savings of $130,000 annually in finance charges, which may even justify hiring an additional employee in the Collections Department.

Income Taxes

The dark side of having fewer bad debts is higher income taxes. A good tax accountant or attorney will help decrease the company's cash flowing to the government. Not only can he or she help save cash on current taxes due, the accountant should also be consulted when the company plans future projects. The use of accelerated depreciation, deferred compensation for executives, timing of charitable contributions, capitalization policy, lease/buy decisions, and financing are all areas that a qualified tax consultant can work to positively affect cash flow. Even though most tax savings are a matter of timing and are merely postponing payment to the government (i.e., dollars saved now will be paid later), it is still better to hold the cash for a few years than to pay the government immediately. Have the tax consultant consider tax implications of all major business decisions. More detail on tax-saving ideas can be found in Chapter 15.

Managing Accounts Payable

Like taxes, accounts payable is an area in which the business will not normally save cash permanently, but will temporarily defer payment. (Some discounts are available for early payment; however, the value of this discount is a function of the company's cost of capital.) Lengthening payment timing can cut both ways by straining vendor relations. A vendor that has been getting paid slowly may be reluctant to fill the next order for tape stock, cable, or a tape machine. Many times, market conditions dictate how long a payment can be delayed.

Of course, it is good business practice not to pay a bill before it is due, but delaying payment too long can backfire. Be careful: short-term cash flow gains can significantly impact long-term operations (bad credit referrals, cash-in-advance terms, etc.).

Fixed Asset Purchases

When considering whether to lease or buy a fixed asset, always be sure to perform a computation to determine the interest rate included in the lease payment. (Although the actual formula is a little more complicated, this is basically the percentage difference between the sum of all lease payments *plus* the cost to purchase the asset at the end of the lease and the cash price for the asset.) If the rate is higher than the interest rate on the company's credit lines, then it is generally more beneficial to purchase the asset outright.

Cash Planning

Do not forget cash planning when trying to enhance cash flow. Planning cash needs in advance can help shift excess cash from low-interest or non-interest-bearing accounts to higher-return accounts. Alternatively, if an increase in excess cash is not currently needed, the company may decide to pay down debt carried at a high interest rate. Other methods of getting the most from cash are the use of a secure lockbox for collections of receivables, and zero-balance disbursing accounts. A lockbox is a bank service in which customer (advertiser, subscriber, etc.) payments are sent to a post office box managed by the bank; checks are retrieved, processed, and credited to the company's account more quickly than those sent to the company's office—this reduces the number of days it takes to convert a customer payment into available cash. In a zero-balance system, funds are transferred daily from a central account in an amount equal to the checks presented for payment that day. Consolidating accounts and utilizing a zero-balance system can help reduce the cash balances that would be required by maintaining many accounts.

Many banking institutions provide cash-management software that will carefully track the movement and balances of a company's cash funds. These systems are usually inexpensive, and facilitate full productive use of all funds of a company.

In Conclusion

A few moments spent with your banker, your tax consultant, and/or cash-management experts will pay big dividends when it comes to maximizing your cash flow.

Part II

10 Sarbanes-Oxley and Internal Controls

Mary M. Collins

You're a very successful advertising sales manager who's just taken a job with a new company. Everything is great—the pay and benefits are much better than those of your previous job. The only problem is that every time you do something and don't go exactly by the book—whether it's signing an insertion order before the credit's been approved or turning in an expense report without itemized meal receipts—the business manager comes down on you. She keeps telling you that you have to do it that certain way "because of SOX" (Sarbanes-Oxley), or "because of Section 404." You've heard these terms in the news, but never paid much attention to them. Something tells you that you should now. Mary M. Collins is President and CEO of the Broadcast Cable Financial Management Association (BCFM), and she provides an insightful look at this often exasperating yet powerful piece of legislation.

Introduction

In this chapter, the reader will gain a basic understanding of the Sarbanes-Oxley Act of 2002 and its impact on broadcasting and cable companies. In it, you will find:

- An overview of the key provisions of the act.
- A summary of Sarbanes-Oxley Section 404: Management Assessment of Internal Controls.
- How COSO (to be explained later) has become the de facto framework for certifying internal controls.
- How Section 404 has impacted business in general, and broadcasting and cable companies in particular.
- A look at pending changes to the regulation and how they may impact affected companies.

In the late 1990s and early 2000s, a number of corporate accounting scandals robbed shareholders of their investments. The images of retirees losing their savings, and children with worthless college funds prompted Congress to action. The result was the Public Company Accounting Reform and Investor Protection Act of 2002, commonly known as the Sarbanes-Oxley Act of 2002, or SOX.

This law mandated dramatic changes in the ways that public companies conduct their business and report their financial activities. Divided into 11 titles or subsections, SOX provisions include:

- The confirmation of the Securities and Exchange Commission (SEC) as the ultimate authority for oversight and management of SOX and other U.S. securities legislation.
- The creation of the Public Company Accounting Oversight Board (PCAOB), a new quasi-governmental agency responsible for overseeing, registering, and disciplining public accounting firms as well as for setting audit standards.
- The requirement that the Chief Executive Officer (CEO) and the Chief Financial Officer (CFO) of each affected company certify all periodic reports containing financial statements.
- The requirement for additional real-time disclosures.
- The requirement that all corporations listed on public stock exchanges have an independent audit committee that manages the relationship between the audit firm and the company.
- The specification of auditor independence—banning a company's auditors from performing certain types of work, requiring precertification of all work by the audit committee, and requiring that accounting firms rotate the lead partner and review partner on accounts to ensure that neither accountant fills the same role for more than five years.
- The prohibition of personal loans to any corporate officer or director, with very limited exceptions.
- A strengthening of regulations addressing insider trading—with prohibition on such trades during certain blackout periods, and a requirement that all such trades be reported within two business days.
- Identification of additional actions as securities crimes (such as destroying, altering, or falsifying financial documents), strengthening the penalties for these white-collar crimes, and specifying a new statute of limitations for such crimes.

Section 404: Management Assessment of Internal Controls

The most complex and expensive provision of SOX is Section 404: Management Assessment of Internal Controls. One BCFM member succinctly summed up its impact when she said, "Sarbanes-Oxley is certainly a new career."[1]

Sarbanes-Oxley uses the term "issuer" to describe any company covered by the legislation; this term comes from the Securities Exchange Act of 1934. Broadly stated, SOX defines "issuer" as any company—whether headquartered in the United States or elsewhere—whose securities are registered under the Securities Exchange Act and which company is required to file reports under Section 15(d) of that act.

SOX Section 404 requires each issuer to include in its annual report an internal controls report that includes three fundamental components:

1. A statement asserting that the corporate management is responsible for establishing and maintaining adequate internal controls and policies to ensure accurate financial reporting.
2. An assessment of the effectiveness of the internal controls and procedures for ensuring accurate financial reporting, as of the end of the fiscal year for which the report is being issued.
3. An audit, performed by the company's independent auditor, of management's assessment of the effectiveness of the company's internal controls over financial reporting.

It is interesting to note here that the congressional committee report that accompanies the bill, a report that is intended to explain the committee's legislative intent, says "... the committee does not intend that the auditor's evaluation will be the subject of a separate engagement." That is to say, Congress did not believe that a company would incur any additional expense for this assessment. Ah, the naïveté of our elected officials.

As stated above, the SEC is ultimately responsible for issuing the regulations for enforcing SOX. In the case of Section 404, the first assessments of internal controls were due before the SEC issued final rules about how companies were to comply with internal controls reporting. Companies and their auditors identified a

---

1. Deborah Cowan, vice president of finance for Radio One, Inc., as quoted in *The Financial Manager,* January–February 2005.

voluntary U.S. standard known as COSO as the only standard that seemed to meet the requirements of the act.

Formed in 1985, COSO is a framework developed by the Committee of Sponsoring Organizations of the Treadway Commission. (Although the SEC has not mandated COSO as of this writing, it has stated that COSO's control standards satisfy the agency's criteria for an acceptable framework for evaluation.) This committee is sponsored and funded by five main professional organizations: the American Institute of Certified Public Accountants (AICPA), the American Accounting Association (AAA), the Financial Executives Institute (FEI), the Institute of Internal Auditors (IIA), and the Institute of Management Accountants (IMA). COSO's main objective is to identify factors that cause fraudulent financial reporting and to make recommendations to reduce their incidence. To do so, the committee established a definition of internal controls as "... a process which provides reasonable assurance that an organization will achieve its objectives for effective and efficient operations, for reliable financial reporting, and for compliance with applicable laws and regulations."

The committee also established criteria against which companies can assess their control systems. As stated in 2003, COSO had five components:

1. **Control environment**—Encompasses such factors as integrity, ethical values, management's operating style, systems for delegating authority, and processes for managing and developing people within the organization.

2. **Risk assessment**—Begins with establishing objectives and identifying and analyzing relevant risks that are prerequisites for determining how to manage the risks.

3. **Control activities**—The policies and procedures used to ensure that management's directives are carried out and that help to ensure that actions are taken to address risks to achieving the established objectives at all levels of the organization. (This includes such things as authorizations, verifications, reconciliations, segregation of duties, etc.)

4. **Information and communication**—Includes both the information systems that produce reports and the procedures that ensure that information flows up, down, and across an organization and to external parties.

5. **Monitoring**—Procedures to monitor internal controls systems, to report deficiencies, and to take corrective action.

Although this sounds straightforward, keep in mind that management must certify internal controls for every *significant* financial account and process. An *account* is considered significant when it can contain errors of some importance and/or if management thinks it should be evaluated. A *process* is considered significant when it covers a major class of transactions, affects significant accounts (or groups of accounts), covers significant volume of a company's activity, and/or impacts a significant account on the organization's general ledger. One public accounting firm has gone so far as to say that every account included on an issuer's external financial statements is significant.

IT (information technology) processes or any procedure that requires the manual movement of data from one person, department, or software program to another requires particular scrutiny. Transactions between related entities (e.g., companies that own both radio stations and television stations, and/or cable systems as well as programming divisions; and companies that manage both businesses that sell advertising time and those that insert advertisements into programming) must be assessed carefully. If the company outsources significant parts of its business or relies heavily on outside vendors, these outside sources, too, must be evaluated.

Examples of affected accounts and processes in broadcasting and cable span the entire operation of the businesses. They include:

- Advertising sales.
- Accounts receivables.
- Capital assets.
- Capital projects.
- Cash management.
- Credit policies.
- Disbursements.
- Financial reporting.
- Information technology systems.
- Inventory management.
- Order entry.
- Payroll.
- Program agreements.
- Nontraditional revenue.

- Sales commissions.
- Special offers.
- Subscription sales.
- Subscriber authorization.
- Subscriber reporting.
- Talent agreements.

For each identified account or process, management must review, evaluate, test, and document using COSO (or another framework that meets the SEC's criteria) to ensure that significant controls and procedures are in place to prevent mistakes or misconduct. Documentation must provide *reasonable* support for the design and conclusion, addressing each of the components outlined in the approved framework.

Given the scope of this certification, many companies have hired consultants to help complete the task. Before SOX, companies would likely have turned to their own auditors for help. Because this is prohibited under the act, those looking for assistance have to engage either another audit firm or qualified group.

The Next Step

Once management has completed the internal certification, the process must also be certified by the issuer's external auditors. Using statistical sampling, each account and process is tested for *deficiencies*. The PCAOB defines "deficiencies" as conditions that, in the normal course of performing the assigned function, cannot prevent or detect financial misstatements on a timely basis. Deficiencies are divided into three levels:

1. **Inconsequential deficiency**—Taken alone, the condition will have a negligible or inconsequential impact on an issuer's external financial statements. Although two or more deficiencies may be considered "inconsequential" individually, in combination, they may reach a level of a "significant" deficiency.
2. **Significant deficiency**—A condition or conditions that have more than a remote likelihood of adversely impacting an issuer's ability to report reliable information on external financial statements.
3. **Material weakness**—A condition in which there is a more than a remote likelihood of material misstatements in an issuer's external financial reports that will not be prevented or detected.

Despite the SOX committee's initial assumption that "... the auditor's evaluation [not] be the subject of a separate engagement," and thus add significant expense, the preparation of the internal controls report has required and will require substantial issuer resources, both in time and in dollars. Estimates of the initial cost of compliance vary widely—2004 estimates ranged from $3 million to $16 million, and more than 30,000 hours per company. It is for this reason that a number of public company boards have opted to take their companies *private*.

Where Do We Go from Here?

As of this writing, two classes of issuers have yet to comply with the reporting requirements of Section 404. They are non-accelerated filers (generally smaller companies with less than $75 million in nonaffiliated market capitalization) and newly public companies (those companies that have not yet filed an annual report with the SEC). Hearing the initial complaints and the ongoing concerns about the cost of compliance, the SEC adopted revised rules for these companies on December 15, 2006. Under these rules, management's reporting requirements for non-accelerated filers become effective for fiscal years ending on December 15, 2007, or later. The auditor's reporting requirements for these filers are effective on December 15, 2008, or later. As for newly public companies, the auditor's reporting requirements become effective with their second annual report.

There are also indications that the PCAOB is looking at ways to make issuers' annual internal controls evaluations more efficient and cost effective. In the works is the PCAOB's revision or replacement of Auditing Standard No. 2, the standard that audit firms use to attest to an issuer's internal controls. Companies hope that the revised standard will focus auditors on material weaknesses in internal controls—these have the highest likelihood of resulting in material misstatements on an issuer's financial statements.

Also to be evaluated is the impact of COSO's 2004 Enterprise Risk Management—Integrated Framework, which expanded the committee's framework to eight components from the five cited in this chapter.

Finally, it will be some time before investors and issuers can accurately complete a cost-benefit analysis of SOX, and particularly of Section 404. Do the benefits of the legislation (and its subsequent revisions) outweigh the cost of implementation and certification? Did SOX meet its objective of restoring lost investor confidence? Does it prevent the abuses it was designed to correct? Or, will history prove the truth of the words spoken by Cox CFO Jimmy Hayes at BCFM's 2004 Annual Conference in Atlanta: "The best controls won't help a company that has lost its integrity"?

Bibliography

AICPA. "Sarbanes-Oxley." AICPA. http://cpcaf.aicpa.org/Resources/Sarbanes+Oxley/ (accessed December 29, 2006).

———. "Sarbanes-Oxley—The Basics." AICPA. http://cpcaf.aicpa.org/Resources/Sarbanes+Oxley/Sarbanes-Oxley+ – +The+Basics.htm (accessed December 29, 2006).

CPE Inc. Online. "COSO & Creating Anti-Fraud Programs Under SOX." CPE. http://www.cpeonline.com/cpenew/courset.asp?topic1=A246 (accessed December 29, 2006).

———. "Sarbanes-Oxley Act of 2002." CPE. http://www.cpeonline.com/cpenew/sarox.asp (accessed December 29, 2006).

Ernst & Young. "Legislative Summary of the Sarbanes-Oxley Act of 2002," Ernst & Young, July 29, 2002.

Infinity Broadcasting. "Station SOX Compliance Manual." Infinity Broadcasting, April 26, 2006.

KPMG 404 Institute. "SEC and PCAOB Proposals Regarding Internal Control Over Financial Reporting." KPMG 401 Institute. http://www.404institute.com/docs/SEC_PCAOB_Ann_122106.htm (accessed December 22, 2006).

Maltese, Evan (E&Y LLP), Lawrence Wills (Granite Broadcasting Company), and Chris Pimentel (Entercomm Communications). "BCFM Distance Learning—Section 404 of the Sarbanes-Oxley Act." November 4, 2003.

PricewaterhouseCoopers. "Management's Responsibility for Assessing the Effectiveness of Internal Control Over Financial Reporting Under Section 404 of the Sarbanes-Oxley Act." PricewaterhouseCoopers, December 2003.

PricewaterhouseCoopers CFOdirect Network. "Breaking News: SEC Provides Additional Sarbanes-Oxley Section 404 Deferrals for Smaller Public Companies and Newly Public Companies." Pricewaterhouse Coopers CFOdirect Network. http://www.cfodirect.pwc.com/CFODirectWeb/Controller.jpf?ContentCode=AALN-6WHUFB&SecNavCode=USAS-6BG36W&ContentType=Content (accessed December 15, 2006).

SC&H Group LLC. "The Bottom Line: Alerts! Breaking News—SEC Finalizes 404 Reporting Dates for Non-Accelerated Filers." SC&H Group. http://scandh.com/FromSCandH/404_Filing_Dates.html (accessed December 20, 2006).

Schley, Stewart. "Rocked by SOX." *The Financial Manager,* January–February 2005.

SOX-online.com. "Sarbanes-Oxley Act Basics." SOX-online. http://www.sox-online.com/basics.html (accessed December 30, 2006).

Wikipedia. "Committee of Sponsoring Organizations of the Treadway Commission." Wikipedia. http://en.wikipedia.org/wiki/Committee_of_Sponsoring_Organizations_of_the_Treadway_Commission (accessed December 30, 2006).

———. "Sarbanes-Oxley Act." Wikipedia. http://en.wikipedia.org/wiki/Sarbanes-Oxley_Act (accessed December 29, 2006).

11 Strategic Planning and Budgeting

Laura James and Andrew Kober

Many companies turn to strategic planning when they are in trouble, when growth is flat or falling and expenses are on the rise. This may be the worst time to begin the process. In crisis situations, money, staffing, everything is tight, so it's hard to find the resources needed to support new projects. Change cannot happen overnight. It requires both time and money. In this chapter, Laura James—Senior Vice President of Finance for Lincoln Financial Media, which owns 3 television stations in major Southeastern markets, and 18 radio station in Top 50 markets across the country—and Andy Kober—Senior Vice President and Controller for Bresnan Communications, which operates cable systems in nonurban U.S. markets—explain the importance of strategic planning and provide specific steps to help media companies develop meaningful plans for their businesses.

Introduction

The discipline of developing plans and measuring progress toward established goals will be increasingly important as media and telecommunications companies converge in the coming years. Neither the cable industry nor the broadcasting industry can afford to bet their future on the status quo. Most traditional broadcasters and cable companies are asking themselves hard questions about ways to sustain and increase the value of their properties. Identifying the strategies that will keep these businesses vibrant and thriving is the key to successful planning. Quantifying and measuring the strategies is a separate but very important step toward achieving goals.

Much of the financial process looks in the rearview mirror to determine how the business performed, and therefore a financial review can be of limited interest to nonfinancial managers because they already have a general sense of how they performed. *A strategic plan looks forward, not backward.* It is the process of determining a company's long-term goals and then identifying the best approach for achieving those goals. This forward-looking planning and budgeting process gives managers an opportunity to map out their future and communicate those plans to all parties within the organization.

To be meaningful, a strategic plan must be a dynamic document. A company that goes through the exercise to create a well-written plan and then puts it on the shelf until the next year is wasting its resources. An effective plan that has been developed using input from all company departments provides the yardstick against which all employees can be measured.

Strategic Planning

Developing a strategic plan is the first step for any business to drive forward financially. *Basic planning* is something that happens naturally, both formally and informally throughout the fiscal year. Strategic planning is the formal process of working through a defined set of steps that concludes with a master plan that is incorporated into a written document. Industry business managers already have an idea about much of what they would like to incorporate into a strategic plan. The process of formally documenting the strategic plan brings together many of the goals of the company's leaders, organizing their thoughts and putting pencil to paper so that the key leaders/managers are all "on the same page." The process of determining goals, evaluating business options, and eventually agreeing on a plan can be far more important than the final written product.

One aspect of strategic plan design that is sometimes overlooked is *identification of contributors*. Who should participate in the planning process? Although design of the planning and budgeting tools can be delegated to appropriately qualified staff members, senior management and department heads must be engaged in the process from beginning to end, or the plan risks failure from the outset due to lack of buy-in from those responsible for its execution. A strategic plan that includes inputs from all parties critical to its implementation will be much more successful than one developed in a vacuum.

Designing a planning methodology, the approach to the process, is an important step in both strategic planning and budgeting. Some public and private companies have a Planning Department that sets deadlines and defines dates, deadlines, and formats for both the strategic and budgeting phases of the planning process. This Planning Department can also provide a framework or tool that will walk participants through the exercises and result in formal documentation. It is incumbent upon the business that lacks the assistance of a formal Planning Department to take the time to design a framework and establish timelines that will achieve the same goal. Determination of methodology and design of the expected outcome should be an independent step apart from the exercise of completing the documentation. The shortcut of figuring it out as you go along can lead to critical oversights.

The first step in planning must be an honest evaluation of the company as it stands today. What are its strengths? Where are its weaknesses? How about opportunities and threats? Often called a SWOT analysis (Strengths, Weaknesses, Opportunities, Threats), this exercise helps company management really understand the current situation and the environment in which it is operating. A good SWOT analysis will thoughtfully consider the pressures on the business that will need to be addressed during the planning cycle, such as:

- Overall economic outlook.
- Market-specific economic outlook.
- Competitors.
- Emerging technologies.
- Legal and regulatory requirements.
- Demographic shifts.

The first cut at these reviews should be high level without drilling too deeply into specific details. This overview will lead to *identification of emerging issues* specific to the business. Those emerging issues will be the foundation for setting corporate goals.

With the SWOT analysis completed, the next step is for participants to decide where the company should be at the end of the planning period. This may be easier said than done. But it's crucial to get consensus on the vision before moving on to the next step.

Once planning participants agree on the vision, it's time to begin a "gap" analysis. The question to be answered is, "What is the gap between where the company is today and where the planners want it to be at the end of the planning period?" A media company's assessment may conclude that it is short on technology; that its technology is old; that it is being beaten in a specific day part; that it is not meeting the needs of its viewers, its advertisers, and/or its subscribers; or any of a number of other things.

The planning cycle can be any length of time that makes sense for the specific business, typically between one and five years. One year is a minimum, and may be too short term for any business because it can result in a focus that is too narrow and tactical. Three years is the average planning cycle for most companies, at least for the purpose of identifying strategies. Anything beyond three years becomes so distant that participants have a hard time relating to the outlying years. The planning cycle for the strategic plan does not have to be identical with the cycle for budgets or capital expenditure planning.

Strategies flow directly from the gap analysis. They are the long-term action plans intended to move the company to achieve each of the identified goals. One can say that strategies become the road map to move the company from the present to the future. The next step in the strategic-planning process is to *prioritize the business strategies* and assess the cost/benefit of each. These exercises are in lockstep with each other, and can result in reevaluation of previous decisions made during this process.

With strategies identified and prioritized, planners need to set intermediate steps on the way to accomplishing these strategies. These tactics are the mile markers in the strategic road map. It's important to look at the resources needed to complete the tactics assigned to each strategy. If achieving the goals includes people, research, consultants, hardware, and/or other resources, these must be included in the tactics and the budget.

For each strategy, planners should identify key drivers and how they can be measured. When these tactics and their measurements are assigned as responsibilities coupled with deadlines, managers can be held accountable for completing them. These can and should be reviewed at regular intervals to ensure that the company is following the road map it set for itself.

Failure to relate to the strategies can have the dangerous consequence of distancing managers from ownership of the plan, and can result in subsequent disregard for the strategies developed when making tactical decisions. For example, if the goal is to update the company's technological infrastructure, tactics may include spending heavily on capital equipment. A manager not vested in the process may delay capital spending to improve cash flow. In this instance, the results may look good in the short term, but the decision will negatively impact the company's ability to achieve its technology goal.

A strategic plan provides a high-level view of the company's activities for the upcoming planning cycle. The budget quantifies the dollars-and-cents impact of tactics that will be employed to achieve the strategic goals.

Budgeting

Once the strategic plan is drafted, it should be rolled into the budget. The key here is making the tough choices to redirect resources to fund the projects identified in the plan. Usually a detailed budget is developed for the first year of the planning cycle. This budget will project revenues and expenses down to the general ledger line-item level. Outlying years may not require this level of detail, but this is usually determined by the specific needs of the budgeting entity, including upstream reporting requirements.

Due to the cyclical nature of some components of revenue (e.g., advertising revenue and the effect of seasonality on some geographic locations), it is preferable to show the budget month by month. This provides an ongoing measure of strategic performance, and can alert management early on that business plan goals may be in jeopardy. Many companies, particularly those that give guidance to analysts and have quarterly reporting requirements to the Securities and Exchange Commission (SEC) and to their lenders, need budgets subtotaled by quarter.

Budgeting for revenues and revenue-specific expenses should be a top-down exercise *based on the strategic plan*. Revenues from traditional lines of business should reflect the assumptions adopted in the strategic plan for growth of the markets, combined with the specific goals for the business's performance within the market. Additional revenues and revenue categories should be included for new goals identified in the strategic plan.

Expenses should be treated in the same manner. Those for established lines of business should reflect the assumptions used in the strategic plan (e.g., if, as suggested above, the company assumes it will have less business from advertising agencies, then agency discounts will be budgeted at something less than historical levels). Expenses for areas under contract (e.g., rent, talent agreements, etc.) are to be included at the rates in those agreements. Budgeted depreciation and amortization expense must include projections for existing assets and budgeted capital expenditures; depreciation and amortization expense are also adjusted for acquisition and disposition of lines of business. And assumptions about costs for new projects should be based on the research done during the strategic-planning process.

There are various approaches to developing the operating expense budget detail, and one is not necessarily preferable to another. Most broadcasters and cable operators ask their general managers and department heads to generate a detailed operating expense budget for their location for each line item. Salary expense and related personnel costs are usually developed on a position-by-position basis using an overall assumption of annual raises or contractual compensation increases, including benefits and taxes. Other line items can be generated either by reference to contracts or by making reasonable assumptions about cost increases and usage variations from previous years.

The first cut of annual departmental budgets generally includes "wish list" items that may or may not be part of the business's overall strategic initiatives. Once they are combined, it is the responsibility of management to determine which of the items are "must-haves," which are "nice-to-haves," and which may need to be deferred or even refused.

Nonoperating expenses such as interest expense are usually budgeted by finance and treasury managers, and are reasonably determinable based on projected debt levels, contractual arrangements, and interest-rate forecasts. Additional financing costs related to execution of strategic plan initiatives should be included as appropriate (planned refinancing, recapitalization, and other financing initiatives).

At the end of the process, the question to be asked about the budget is: Do the numbers make sense? Do they reflect the business initiatives identified during the strategic-planning exercise?

Subsequent Measurement

Although development of a budget can be a useful exercise on its own, it, like the strategic plan, must also be a dynamic document. The business derives maximum benefit by applying the discipline of subsequently comparing actual results to the plan. Variances, both positive and negative, should be explained (preferably in writing) and understood. The measurement period should be frequent enough to allow for corrective action to be taken for correctable misses and to take advantage of unforeseen positive circumstances as they occur.

Variance reports show the various differences (in dollars and/or percentages) between budget and actual for both revenues and expenses. They are the accepted tool for comparing plans to actual results. These reports should align with levels of budget development (i.e., monthly, quarterly, annually, etc.), and should be prepared at least as frequently as the timelines outlined in the budget (typically monthly). General managers and department heads with responsibility for developing revenue and operating expense budgets should subsequently be responsible for generating variance reports for those revenues and expenses.

Capital Expenditure Budgeting

An important component in developing a complete business plan is capital expenditure budgeting (cap ex). To manage capital expenditures, all broadcast and cable television businesses should maintain a three- to five-year capital plan for replacement of equipment, adopting new technologies, and fostering business expansion. These proposed expenditures must be reviewed carefully for urgency, timing, and alignment with the larger strategic initiatives. Management should prioritize the projects when the strategic plan is prepared in order to make intelligent decisions about which capital projects to include in the current year budget, based on predetermined rules regarding breakeven thresholds, payback periods, and discounted cash flow analysis. For more information on these subjects, see Chapter 7.

In Conclusion

Strategic budgeting and planning, which may seem like a laborious and overly complex process, is an essential key to business success. The time to begin planning is when the company is healthy and growing because that is when it will have ready access to the resources necessary to support the plan. Planning and budgeting activities provide a critical road map in which a company identifies and prioritizes strategies for growth and improvement. An effective plan includes both narrative and statistical (measurable) components. The plan must also include examination and review of proposed capital expenditures. When those who will ultimately be responsible for carrying out the plan are involved from the inception, their buy-in will provide internal momentum to drive the business forward. Finally, it is not enough to simply write a plan. Successful companies measure employee performance against the strategies and tactics in the plan. Subsequent comparison and explanation of variances between the budget and strategic plan and the actual results help companies understand where the plan can and should be modified.

12 Credit and Collection Administration

Peter F. Szabo — Revised by C. Robin Szabo

Every advertising-supported media business extends credit to a greater or lesser degree. These businesses also have to collect funds due. They're not alone. Ever since humans came up with the concept of business, they have faced the challenges of extending credit and collecting when the debt comes due. This chapter, originally written for the first edition of this book by media collections pioneer Peter F. Szabo, addresses the many opportunities and pitfalls in credit and collection. C. Robin Szabo, president of Szabo Associates, revised the chapter for this edition.

Introduction

Every broadcast and cable company has a credit policy, though it may be unwritten. If a prospective advertiser or agency has been given payment terms, has been refused credit, or has been told that the company cannot accept its particular type of advertising, then a credit policy exists. Many credit policies remain informal and unwritten guidelines communicated by word of mouth to advertisers and agencies. However, because unwritten credit policies lack uniformity, conflicting information may be given from the same broadcast or cable company. Worse yet, advertisers and agencies may be treated differently even though no difference exists in credit qualifications.

Well-articulated credit and collection policies (a) minimize the risk of unfair credit decisions (and subsequent discrimination lawsuits), (b) help to establish a positive reputation within the advertising community, and, most importantly, (c) avoid costly payment delays that result from sloppy credit and collection procedures.

Have a Written Credit Policy

When disputes do arise, a written policy clearly establishes responsibility for payment. If your company's clearly stated terms are 30 days, and payment has not been received within 45 days, you rightfully can begin to enforce your collection policy and trim the cost of carrying overdue debt.

Because a credit policy will benefit the Accounts Receivable and Sales Departments, representatives from both areas should be involved in the development of the company's written policy. The credit manager should begin by developing a general outline of requirements and procedures, whereas the general sales manager should become involved in developing more-detailed points of policy that represent both departments' interests.

Policy Objectives

The real challenge in creating a policy is balancing *credit extension* against *collection operations*. For example, one company might choose to balance a liberal credit-extension policy with a conservative collection approach. Another may opt for a tighter extension policy, lessening the need for strict collection procedures. And yet another may strive to achieve an equal balance between the two. In any case, the relative emphasis should be determined at the outset.

When two departments, such as Sales and Finance, with differing objectives are affected, achieving agreement on this emphasis can be difficult. Credit objectives generally include minimizing bad debt while maximizing cash flow. Advertising Departments generally favor guidelines that broaden, or at least will not restrict, sales efforts.

Credit guidelines might also address customer relations. An effective policy should embrace the principles of *fairness, firmness, courtesy,* and *consistency* in relations with customers. And, of course, assignment of the authority to administer the credit policy should be included.

Payment Terms

As part of an overall policy, payment terms should be described in detail: invoice dates; mailing dates; the number of days allowed for payment; procedures for collecting accounts, including when to send letters and make telephone calls on overdue accounts; when to send accounts to a third-party collection agency; and when accounts are to be written off. Samples of collection letters should also be included.

Once a credit and collection policy is in place, review it periodically. One way to ensure that policies always reflect the company's current approach is to make a habit of reviewing and updating the credit policy each time the company changes its advertising rate card.

Upon completion, the policy is typically made part of a standard operating procedure (SOP) manual. That manual should be distributed to the general manager, sales manager, credit manager, and account executives for administration.

Fraudulent Advertising

The direct and indirect costs of fraudulent advertising can be debilitating to a broadcast or cable company. The direct cost is, of course, lost dollars from unpaid advertising. The indirect cost is a loss of reputation and audience, because defrauded viewers may associate the broadcast or cable company with a negative experience.

It is unlikely that an established business will risk the negative attention associated with fraud. Most such advertising is attempted by new businesses or individuals, or by an advertiser that has previously made similar attempts.

Advertising standards, which set the guidelines for acceptability, are essential to help protect against fraudulent advertising. Management sets the standards, which can vary widely among companies and are largely responsible for the company's reputation among viewers.

Some standards are easy to formulate. Advertising that is clearly in violation of the law should, of course, be unacceptable. For advertising that is questionable, but not obviously illegal, the advertising and credit staffs must be given clear guidelines for determining its acceptability.

Advertising and credit staffers should be formally trained in the application of these standards and guidelines. Training for current employees may take the form of dissemination and discussion of changes or new policies. For new employees, a thorough briefing and understanding of the policies before beginning work is important.

The advertising staff is usually the first contact with prospective advertisers and agencies, and should bring questionable commercials to the attention of their managers. If the manager is unable to determine the advertisement's suitability, the matter should then be taken before the Standards Committee.

Prospective advertising that makes it past the advertising staff should then be subjected to scrutiny by credit staff trained in advertising standards. Credit staff can challenge a questionable ad on issues of content and billing, and submit requests for additional documentation to support its legitimacy.

Credit

Even the most legitimate and responsible advertisements can be very expensive for a station when submitted by an advertiser that is not creditworthy. Among the numerous costly consequences are the expense of the sales representative's efforts to get the order, the copy and production personnel to produce the commercial, and the traffic personnel to schedule the commercial, perhaps in time slots that other creditworthy advertisers and agencies may prefer. Additional expenses may include credit personnel to bill, issue aging reports and make collection efforts, and the potential of court fees to try to collect payment.

Many man-hours of company expense can be avoided by thoroughly checking a prospective advertiser's or agency's creditworthiness at the sales order stage. For all new accounts, the general manager and sales manager should demand signed and complete sales orders, credit applications, and payment agreements. This is unnecessary for clients that have a satisfactory track record, or that do business on a cash-in-advance (CIA) basis.

In fact, certain types of accounts should always require prepayment. Examples are accounts involved in bankruptcy proceedings or going-out-of-business sales, concerts and one-time events, most restaurants, and accounts for which creditworthiness cannot clearly be established. Ample time should be allowed for CIA payments to clear the bank before the schedule begins to air.

Political candidates should be covered by distinct and separate guidelines for campaign advertisements because their interest in advertising ends with the election (which may also be when the office that ordered the advertising is closed). For example, most media companies insist on cash in advance for all such advertising. Also, be certain that all political advertisements comply with current Federal Communications Commission (FCC) regulations.

Information Sources

As one of the first forms of media to be used by new advertisers, the local newspaper is a good source of credit information. Local or distant-market broadcast and cable companies that the advertiser or its agency has used are always valuable sources. Local credit bureaus and credit-information providers, such as D&B (formerly known as Dun & Bradstreet) and Experian, although not media-specific, can be utilized as well. However, media-specific information can be obtained from the Broadcast Cable Credit Association (BCCA). The BCCA Credit Reporting Service contains information on more than 25,000 agencies and advertisers, and will perform credit inquires if the information sought is not in its files. Additionally, information to supplement the reference and credit report findings can be obtained through various online sources, such as the agency or advertiser web sites, EDGAR (the SEC's Electronic Data Gathering, Analysis, and Retrieval system), Google, Hoover's, LexisNexis, and Yahoo! Finance.

In order that a timely credit decision can be made, all credit references and banks should be contacted, preferably by telephone, within one day of receipt of the credit application. Making these calls and quickly completing the credit evaluation will allow adequate time to receive a CIA payment or a personal guarantee from the business owner if the investigation turns out to be unsatisfactory.

Credit updates should be completed annually for all active agencies and advertisers, and should be conducted more frequently for those accounts that (a) have a history of delay payments, (b) have issued checks drawn against insufficient funds, or (c) have used questionable credit practices.

Liability

It is critical to determine ultimate payment liability in advance. The broadcast or cable company's liability position must be stated in the sales order under "terms and conditions." There are four liability positions currently in use by media:

1. **Sole Liability** (agency liable).
2. **Advertiser Liability** (advertiser liable). Sales orders should be signed by the advertiser, who should be named on the "Bill to:" line of the invoice.
3. **Dual, or Joint and Several** (both agency and advertiser liable until broadcast or cable company is paid). This is the position recommended by BCFM. Assuming the proper credit checks and notification, this clause protects the company if either party fails to pay.
4. **Sequential Liability** (agency liable to the extent that the advertiser is paid by agency).

Payment Agreements

A payment agreement confirms in writing that the advertiser or agency has agreed to make payments by certain dates. Ordinarily, payment terms are 30 days. If the advertiser or agency wants different terms of payment, these exceptions should be stated at the time the order is written. The broadcast or cable company can then choose to accept or reject these terms of payment.

If the prospective advertiser or agency is a *corporation* with credit determined to be *unsatisfactory*, the broadcast or cable company may choose to accept the advertising with either a CIA payment or a personal guarantee. A personal guarantee is meaningful only if the owner (stockholder) has the resources to pay. The guarantee must be signed by the owner as an *individual,* and not as an officer of the corporation.

Sales and Credit

Conflicts of interest between media Credit and Sales Departments are inevitable, and sometimes can erupt into angry confrontations. When dealing with questionable accounts, the credit unit that is responsible for minimizing past-due receivables and bad debt losses will insist on prepayment terms. At the same time, sales representatives, focused on meeting quotas and increasing commissions, may try to circumvent the system in order to salvage existing accounts and sell new ones. Reconciling the divergent interests of these two groups can seem like a formidable, if not impossible, task; however, by recognizing and focusing on the *shared objective*—keeping credit losses below a certain percentage of sales—instead of conflicting ones, the Sales and Credit Departments can form a profitable partnership. At the heart of an effective sales/credit team effort is good communication. Informal communication between departments as issues arise, as well as formal communication in regularly scheduled meetings, prevents potential problems from becoming reality.

Formal credit meetings should be held at least once a month, with the participation of the sales manager, other sales personnel as appropriate, the business manager, and other management personnel as requested by the broadcast or cable general manager. These meetings should encourage understanding of the reasons behind credit policies in general, as well as decisions about specific accounts. Sales representatives may not realize that the broadcast or cable company suffers increased costs until payment for advertising is received; that is, *unpaid advertising is more costly than unsold advertising.* Credit staffers can also broaden their perspective by trying to understand the pressures that sales representatives face each month; these pressures increase greatly in economic downturns.

Communication is particularly critical when implementing changes to the credit program. If the credit function has been lax and disorganized, sales personnel may balk at policies that seem radical in comparison to previous credit operations. If involved in the process, however, salespeople are more likely to accept change.

Of course, their acceptance of the policy will be limited if the interests of the Sales Department are not being served sufficiently. “Enlightened self-interest” will most effectively benefit both individual and company; therefore, incentives should be included to further motivate sales representatives to cooperate fully with the Credit Department. These incentives should be designed to focus the Sales Department on the new policy or policies. As such, the incentives will be most effective if they are outside of normal compensation. They could include a bonus, a contest, prizes, or something else; local sales management will know how best to get the attention of the sales team.

Sales and Credit Departments can cooperate to maximize collected revenue in a number of ways, such as making sure that credit applications on new accounts contain complete information. In return, the Credit Department should reciprocate by quickly approving an advertiser’s credit upon receipt of the proper credit information to assist the sales representative in making the sale. When required by credit policy or when appropriate to the situation, salespeople should collect prepayments.

Because the sales representative maintains continuing personal contact with advertisers and agencies, he or she is in the best position to become aware early of management or cash flow problems that could make collection difficult. Sales representatives should be alert to such situations, and report the information to the Credit Department immediately. Additionally, sales representatives should review accounts receivable reports on a regular basis. They should formulate a plan with the Credit Department to collect on delinquent accounts before or at the same time as soliciting new advertising from a troublesome client.

Management may also choose to withhold commissions on sales until the money is collected. Although this approach successfully responds to a typical problem among sales personnel—to concentrate on the immediate sale without concern for its long-term collectability—withholding commissions separates in time the sale from the reward. But as any student of psychology knows, timely reward for effort is a great motivator. Management might instead opt to advance commissions to the sales rep the month sales are made. If a sale is not collectible, the commission advance can then be charged back.

Of course, the “carrot” of getting a bigger commission check will always win out over the “stick” of a subsequent chargeback. With adequate controls, though, the chargeback method can make overzealous sales employees aware of the cost of poor credit procedures.

Credit applications on new accounts should be submitted and approved by the Credit Department before the advertisement is aired. The customer’s credit limit and payment history should be reviewed prior to any advertisement’s running.

The Business Office must review each credit application and perform appropriate credit checks to assess client risk. Credit applications should be processed on a timely basis, and notification should be furnished to the sales manager. The general manager's approval should be required in order to override the Business Department's rejection of client credit.

Collections

Even when effective credit policies and procedures are in place and properly implemented, there will be a need for collection efforts in any organization that extends credit. If a broadcast or cable company never has bad debts, it probably means that the firm's credit policies are overly strict and that it is denying itself business.

Reasons for Nonpayment

Some collection efforts result, unfortunately, from errors or omissions on the part of broadcast or cable company personnel. A typical reason for nonpayment is that the sales order was not authorized by the advertiser or agency. Orders must be signed to ensure their validity. Commercial copy and tape not approved prior to the commercial's being aired is another common reason for nonpayment. If the broadcast or cable company produces the commercial, its company personnel must make sure that the advertiser or agency has accepted the final version before putting it on the air. Wrong airtimes or dates are another common reason for nonpayment. It is the responsibility of the Traffic Department to ensure that schedules run as requested unless proper authorizations for change are received. And finally, advertisers and agencies may refuse to pay if invoices and required supporting documents are incorrect or are not received on a timely basis.

Lost billing that results from broadcast or cable company personnel's errors or omissions should be charged to a sales allowances and billing adjustments account instead of to bad debt expense or the revenue account. Although it is lost revenue nonetheless, billing adjustments and allowances segregated in this manner will be more likely to come to the attention of the general manager.

More common than errors by personnel are nonpayment situations in which the broadcast or cable company has acted correctly in all aspects of filling the order and invoicing the advertiser or agency. Success in recovering these slowly paying accounts receivable depends on consistently following structured policy guidelines. These guidelines should clearly indicate to the collection manager what steps should be taken and when, including the criteria to consider turning the account over to a third-party collector.

Collection Etiquette

All stages of the collection process should be conducted in a professional, courteous, and tactful manner. To do otherwise is simply not good business sense, and may expose a broadcast or cable company to legal consequences. Certain types of collection activity are prohibited by law, such as threats of violence or physical harm to a person or property, obscene or profane language, negative statements about a person's character, false or misleading representations of identity or of the debt, and relating false credit information about one person to another. Although it is permissible to give requested credit information to legitimate reporting agencies and parties, the information must be limited to the facts of the experience. Defamation of character, libel, and slander are serious offenses punishable by law.

Legal ramifications of particular collection tactics aside, abusive behavior accomplishes little, and can irreparably damage relations with the advertisers and agencies who intend to pay and to continue to advertise with the broadcast or cable company. In the beginning stage of the collection effort (from 30 to 60 days after the invoice date), the advertiser or agency should be given the benefit of the doubt that the invoice was not received, was misplaced, or that nonpayment was due to a simple oversight. The intention at this point should be twofold. First, motivating the advertiser or agency to pay on prior invoices, and second, maintaining a cordial relationship that encourages the client to continue advertising on the station or cable system. And although the primary purpose of the contact is to collect the overdue amount, an important secondary purpose is to "train" the client to adhere to the agreed-upon payment schedule.

Many advertisers and agencies require only a simple one-time reminder to send payment. Others need to receive several more letters or calls. Each successive letter or phone call should become firmer and create more pressure to pay. For a few clients, the threat of legal action or turning the account over to a third-party collector is necessary in order to obtain payment. One of the most effective approaches to the collection call or letter is to appeal to the client's sense of fairness and pride. If this approach fails to result in payment, an appeal to self-interest—that it would be to their own benefit to pay in order to avoid legal action or third-party collection—is often successful. In any case, persistent and consistent requests for payment will achieve faster results than will casual, erratic efforts.

It is the choice of each broadcast or cable company to determine whether communication by letter, telephone, or a combination of both will best suit its needs. Either can be an effective means of collecting overdue accounts. What matters most is that collection efforts are carried through on a timely basis.

One recommended schedule is as follows:

15 days past due—Send a delinquency notice a few days prior to an account's reaching 30 days past due.

30 days past due—Make personal contacts at least twice prior to the account's reaching 60 days past due. (Earlier contacts may be required for local direct, new, or problem accounts.) Furnish a monthly listing of delinquent accounts to sales personnel, with specific instructions for required assistance.

60 days past due—With the general manager's approval, notify the client that advertising schedules will be taken off the air within 15 days and/or new advertising schedules will not be accepted. Review status of critical accounts receivable with the sales manager and general manager.

75 days past due—Send client a "demand payment letter" or other communication indicating the following: "Unless the delinquent amounts are received within 15 days, the account will be turned over to a collection agency."

90 days past due—Upon approval by the general manager, turn the account over to a collection agency for collection, at which time the amount should be written off. (Check current tax regulations for deductibility of bad debts.)

Collection Letters

The collection letter can be an effective tool in the majority of cases, and particularly with smaller accounts, because a letter takes less time, effort, and money than any other means of collection. Also, a carefully worded letter can be the least offensive collection communication, serving as a "gentle reminder" that will often produce the payment.

A broadcast or cable company should develop its own series of collection letters, taking into consideration the special conditions that exist in the industry and its own credit policies. Two or three variations of the same message can disguise the letter's appearance as a standard form and add flexibility. Guidelines can be drawn up to help credit staff personalize the letters without altering the message.

Collection letters that work best follow several principles for writing effective business letters:

- The writing style is clear and uncomplicated.
- The structure is simple and easy to follow.
- The letter easily fits a single page.

All correspondence should advise the advertiser or agency to ignore the letter if indeed payment has already been sent. Collection letters are an annoyance to clients and can foster bad feelings, particularly if the client has already mailed payment. It is also wise to have an attorney review collection letters before use, in order to ensure compliance with federal and state legislation.

Telephone Collections

Although the approach to telephone collecting is basically the same as with collection letters, there are some significant differences. Collecting by telephone can be effective if it is possible to talk to the person who will authorize payment of the invoice. If that person is never available, credit personnel can make countless phone calls to no avail. The advantage of telephone collecting, however, is that the caller receives immediate feedback on the reasons that the bill has not been paid. Problems can often be resolved in the course of one telephone call.

Telephone collection calls require precall planning to be most effective. The account should be researched thoroughly to make sure that the money has not, in fact, been received, and that the account was billed correctly for the service actually rendered. It is also important to be aware of any previous collection activities—calls, letters, or personal visits—and their results. Finally, the advertiser's or agency's past record of payment should be checked to determine whether a pattern of delinquency exists.

This extra effort in advance should prevent the necessity of having to make repeated calls to the client and gather additional information. Avoiding the embarrassment of being told that payment has already been received, or that the client should not have been billed in the first place, is strong motivation to perform the advance research.

The caller should begin the conversation by first making sure he or she is speaking with the right person. The caller should then identify himself or herself, identify the name of the broadcast or cable company, and state the reason for the call (to determine when payment will be made). An example of an effective, simple, and clear statement is: "Mr. Smith, the amount of $3,000 is outstanding on your account and is 15 days past due. Will you mail a check for $3,000 today?"

After completing this statement, the caller should stop talking and wait for a response. This is the time to listen carefully, to make notes, then to follow with questions that will circumvent excuses (if necessary) and uncover the real reason for nonpayment.

The caller should then state that a commitment is needed now as to when full payment will be made (partial payments might be agreed to eventually, but only if they seem to be the only option). Upon completing the phone call, the caller should send, on the same day, a brief letter confirming the agreement reached on the phone.

Unfortunately, for all its advantages, telephone collecting has one serious shortcoming. It is impossible to physically collect money over the phone. The only way to make sure a telephone collection effort is complete and effective is with prompt and determined follow-up. The follow-up date should allow sufficient time for the check to get through the mail and for transmittal to take place. It should not be so far in the future that the customer senses that collection of the invoice is not a priority.

Final Options

The chances of collecting deteriorate rapidly after 90 days. If, after 120 days and a series of phone calls and letters, the bill is still unpaid, it then becomes necessary to follow through on threats of legal action or third-party collection. It is far better to recover some of the account due than none at all. For that reason, the cost of an attorney or third-party collector should be considered a cost of doing business. Action should be taken quickly, or the risk of losing the entire amount increases.

In Conclusion, Have a Current Credit and Collection Policy, and Keep It Current

A written, formalized, well-thought-out credit and collection policy that includes specific terms of payment and procedures for collecting delinquent accounts is the cornerstone of successful credit and collection administration. Once the policy is in place, communication and cooperation between the Accounts Receivable and the Advertising Sales Departments, as well as timely enforcement of policy procedures, are critical if broadcast and cable companies are to minimize costly payment delays and losses.

13 Trade and Barter Transactions and Related Accounting

Laura Daigle and John Kampfe

When broadcasting was young, operators would frequently make a very nice living without taking in much money. They would trade advertising time for cars, homes, autos, vacations, and food. The Internal Revenue Service (IRS) took a dim view of trade-outs, and, with the backing of the Federal Communications Commission (FCC), a rigorous standard was established. In this chapter, Laura Daigle, Operations Specialist for Clear Channel Communications, and John Kampfe,[1] CFO of Turner Broadcasting System, tell us all we need to know to use trade and barter transactions wisely and legally.

Introduction

The purpose of this chapter is to (a) define trade and barter revenue and expense transactions as they relate to the broadcast and cable industries, and (b) discuss proper recording procedures for these transactions in accordance with Generally Accepted Accounting Principles (GAAP), as well as reviewing (c) internal controls and (d) adequate trade records with supporting documentation.

Trade is a noncash transaction in which a media company receives goods or services from an advertiser in exchange for commercial airtime of similar value.

Barter is a programming transaction in which a client provides a network or station a program that includes commercial spots or sponsorships that the client has sold to a third party. Barter programming may contain some local commercial avails; the network or station retains the revenues from advertising sold to fill these avails.

---

1. The authors would like to acknowledge the work done by Byrne Hopkin and Wayne Frankenfield, whose original version of this chapter was published in the 1993 edition of this handbook.

Trade and barter have economic justification because they provide the media company with items of value without requiring the use of cash. In addition, trade spots typically run in what would otherwise be unsold commercial inventory. Like empty seats on a plane that has just left the gate, unsold commercial inventory has no value once the scheduled airtime has passed. Some advertisers request a combination commercial schedule that is part cash and part trade—in order to make the cash sale, it may be necessary to accept the trade sale.

Barter programming often airs in less-desirable day parts, or on weekends when the airtime may be more difficult for the media company to staff and/or to sell cash advertising spots. Some program agreements combine a combination of cash and barter. In this case, the client both includes spots in the programming and receives cash from the media company.

A media company must carefully review trade and barter transactions to ensure that all costs associated with the trade or barter arrangement are known before finalizing the arrangement. Following is a list of some cash costs of trade and barter transactions that should be considered:

1. **Personnel involvement**—Because trade transactions are not driven by cash, additional approvals and controls are necessary to protect the media company's assets and liabilities. Although the goods and services may have been acquired "for free," the implementation and controls for trade and barter transactions are quite labor intensive. There can also be costs associated with production personnel to prepare the spots or program to air.

2. **Inventory tracking**—Often the assets acquired in a trade transaction consist of hard goods, such as electronics, that must be inventoried and expensed as the goods are used. In order to protect the assets from theft or unauthorized use, the hard goods must be secured and periodic physical inventory taken.

3. **Lost-cash inventory**—As the media company's airtime inventory is sold to cash clients, it may be necessary to forego additional cash sales in order the meet the trade-contract obligation. When a sale to a client is part cash and part trade, the client will pay cash only for the agreed percentage of cash advertising. If the entire schedule is not run as ordered, the client may deduct from the cash schedule if an agreement for make-goods to the schedule is not made.

4. **Uneven trade ratios**—In some trade arrangements, the media company may agree to exchange spots with a greater dollar value (when calculated at full rate or other agreed-upon price) than the fair market value of the goods or services received. Such an agreement discounts the station's inventory, and could make it more difficult to make cash sales at standard rate card prices.

5. **Junk trades**—Trade arrangements can result in acquiring goods for which the media company has little or no use.

6. **Budgeting considerations**—Trade usage is an operating expense, and may reduce available budget for cash transactions.

7. **Sales tax**—Depending upon local tax laws, the goods or services could be subject to sales tax, payable in cash.

Guidelines for accounting for trade and barter arrangements are provided by the Statement of Financial Accounting Standards No. 63 (known as FAS 63), issued in June 1982, and by Accounting Principles Board Opinion No. 29, "Accounting for Nonmonetary Transactions." These have been periodically updated, but the fundamental accounting rules are:

1. All trade and barter should be recorded at the estimated fair value of the goods or services received.
2. Revenue should be recognized in the period in which the spots are broadcast.
3. Expenses should be recognized when the goods or services are received.
4. If goods and/or services are received prior to the broadcast of the airtime, a liability for the advertising time due the client should be recorded. If the spot is broadcast first, a receivable for the goods or services should be reported.

Accounting Principles Board Opinion No. 29, "Accounting for Nonmonetary Transactions," states:

> Accounting for trade and barter transactions should be based on the fair values of the goods and services involved and which is the same basis used in cash transactions. The fair value of the asset or services received should be used to establish the value for the air-time surrendered by the broadcaster.

Fair value of items involved in a trade or barter transaction may be established by referring to quoted market prices, independent appraisals, or other estimates of fair market values. To the extent that one of the parties in a trade or barter transaction could have elected to receive cash instead of goods and services, the amount of cash that could have been received may be evidence of the fair value of the goods or services exchanged.

The valuation of barter programming becomes more challenging when the program is unique and new to the market. Typically, it is appropriate to value the barter programming based on the value of alternate programming in the same time frame. However, if no comparable alternative programming exists at the station, it may be appropriate to value barter programming based on the value of the airtime surrendered. In this instance, care should be taken to value airtime based on similar sales in the same time frame. For example, the highest station rate card would obviously be inappropriate unless the barter program airs in the highest rate card time frame.

Case Study

Example of Trade Transaction in Which Airtime Is Exchanged for an Automobile

Airtime per Rate Card		$40,000
Automobile Sticker Price	$25,000	
Price after 20 Percent Dealer Discount	$20,000	
Dealer Cost	$15,000	

According to the APB's "Accounting for Nonmonetary Transactions," the proper value of the airtime surrendered and the automobile received would be $20,000. This is the fair value of the asset received on a cash basis. The revenue in this example would be recognized when the advertising is broadcast. The automobile would be added to the fixed asset property listing at $20,000, and the media company would depreciate the automobile according to its adopted policy. Implicit to the presumption is that this is an arm's-length transaction—that is, a transaction in which the buyer and seller act independently so that there is no question of conflict of interest. It is also implied that the media company would have elected to receive cash if it had the ability to sell airtime at the rate card amount in excess of the value of the automobile received.

Case Study

Accounting and Journal Entries Associated with a Trade Transaction

For this example, $10,000 of airtime is exchanged for $10,000 of floral arrangements from a florist. Following GAAP, no general ledger entries are made until the florist's advertisements air or the station receives the goods or services. However, many stations record the contract amounts when the contract is agreed and executed.

A. The contract is recorded:

Trade Receivable (asset)	$10,000	
Balance Due Client (liability)		$10,000

B. In the first month of the contract, $2,000 of airtime is broadcast, and $2,000 of floral trade is used:

Accounting for Airtime		
Floral Liability	$2,000	
Trade Revenue		$2,000
Accounting for Trade Usage		
Floral Expense	$2,000	
Trade Receivable		$2,000

The asset account that was set up to show the floral services due is now down to $8,000, and the balance due the client in the liability account is also down to $8,000.

This accounting method applies even when the services used by the media company and the airtime expended do not flow equally. For instance, in the second month of the contract, $2,000 of airtime is broadcast, and $1,000 of floral trade is used.

Accounting for Airtime		
Floral Liability	$2,000	
Trade Revenue		$2,000
Accounting for Trade Usage		
Floral Expense	$1,000	
Trade Receivable		$1,000

At the end of the second month, the asset account that was set up to show the floral services due shows a remaining balance of $7,000, and the liability has a balance of $6,000. In the income statement, on a year-to-date basis, $4,000 of trade revenue has been recognized, and $3,000 of trade-usage expense has been recognized. For a GAAP presentation, there is a net trade receivable of $1,000. This net position at the end of the month should be reported in the balance sheet as an asset, requiring the reclassification of the "Advertising Services Due" balance as an offset to the "Floral Receivable," essentially reflecting collection of a portion of the receivable balance (all but the $1,000 remaining). This demonstrates the effect of timing on the media company's financial statements. At the end of the second month, the asset and liability no longer net to zero, nor do the revenue and expense net to zero. At the completion of the contract, the asset and liability accounts reduce to zero, and the income statement reflects $10,000 of trade revenue and $10,000 of trade-usage expense.

Figure 13.1 provides an example of a form used in tracking trade contracts in an accounting system that combines both trade asset and trade liability on a single form. Each line provides the detail for a single trade contract. The column headed

E. Sample Monthly Trade Agreement Report

Station ____________________

Fiscal Period Ending ______________

			Original Contract Amount		Previous Balance*		Amount Used This Period		Balance Due End of This Period		
Advertiser/Agency (1)	Trade Number	Contract Term: From/To (2)	Time Due Client (3)	Services/ Merch. Due Station (4)	Due Client (5)	Due Station (6)	Client (7)	Station (8)	Client (9)	Station (10)	Type of Merchandise or Service (11)
		Total Page									
		Grand Total									

*Amounts in Columns 5 & 6 Must Agree with Columns 9 & 10 from Last Report

FIGURE 13.1 *Sample Monthly Trade Agreement*
Source: *BCFM Trade and Barter Guideline*—revised 2007.

"Due Client" lists airtime liabilities of the media company, and the column headed "Due Station" lists the media company's assets, which are the goods or services due from the client.

This report becomes a perpetual tracking system for the dollar balance of airtime and trade usage according to the trade contract. It can be used for trade tracked in a single account, or in an accounting system that segregates the trade asset and liability. It should be reconciled monthly and approved by the Business Department.

According to FAS 63, those contracts in which the media company, at the end of the accounting period, has a larger liability than related receivable should be reclassified in a GAAP presentation to the liability section of the balance sheet. The account may be titled, for example, "Advertising Services Due Client."

In a barter transaction, the acquired asset is recorded as "Programming Rights," whereas the income statement would likely include an expense account title "Program Amortization Expense" or a similar title. The income would be titled "Barter Income."

If the asset acquired through trade or barter has a useful life exceeding the media company's policy for expensing an item—a fixed asset, for example—the asset would be depreciated or amortized according to established policy. In the example of the automobile trade above, it is conceivable that all trade revenue could be realized in one year, but the related automobile depreciation would be recognized in more than one year.

Internal Controls Related to Trade Arrangements

Trade contracts require special controls to protect the assets of the media company. As an example, many contracts provide that the traded services be used within a fixed time frame. If the available balance is not used by the expiration date, the receivable must be written off. Although deadlines may increase pressure to use the goods or services, controls must be in place to ensure the proper authorization of their use by station or system personnel. Controls must also be established to ensure that the airtime liability is recorded for advertising performed within the contract parameters, while protecting the media company's available inventory from being used for trade contract obligations that might otherwise be sold to cash customers.

The most important control is to establish a trade policy, the purpose of which is to define the objectives and criteria for entering into trade arrangements. A trade arrangement should improve the media company's overall financial position, not become a drain on sales inventory and employee time.

Case Study

An Example of Points Included in a Trade Policy

- Exchange airtime for client's goods and/or services if the client will not spend cash, if the trade contract is in addition to a cash transaction, or if the airtime cannot otherwise be sold for cash.
- Trade arrangements must be for company use, employees' use, or a combination. If they are for an employee, the goods and services must be specifically approved by management prior to agreement with client.
- Sales commissions paid on trade sales will be recommended by management, with final approval by senior management.
- All trade arrangements must receive written approval of station manager prior to commencement of the airtime schedule or use of the client's services or merchandise. (Specify when additional approvals are needed. For example: uneven trade ratios—that is, trades in which one party receives more value—require regional management approval.)
- All trade transactions must be approved and processed through the accounting records according to GAAP. (Identify level of management review and approval.)
- Trade accounts must be audited according to standard auditing procedures.

This trade policy example sets forth the purpose of the trade arrangement and the deposition of the trade balance, establishes authorization controls, and specifies accounting procedures. The policy also assists all levels of management in their evaluation of trade arrangements.

In a system of internal controls, the key element is the *segregation of duties* insofar as is economically feasible—for example, an employee who benefits from a trade arrangement should not be able to provide final approval for the trade. Figure 13.2 lists the proper division of duties and responsibilities for a typical station.

The trade process at the station, cable system, or cable network usually begins with the sales account executive (AE), because the AE is usually the first point of contact with clients. Trade agreements can be initiated by the station/system/network or by the client, depending on circumstances. Regardless of who begins the negotiations, the essential advantages are the same. Each party wants to reduce cash outlays, either for goods and services or for commercial airtime.

	Salesperson	Sales Mgr	Credit Mgr	Controller	Acctg Clerk 1	Trade Clerk/GM Sec	Acctg Clerk 2	GM	Division Mgr
1) Initiates barter	■								
2) Approves barter transaction		■	■	■				■	■
3) Records transaction in general					■				
4) Maintains custodianship of barter merchandise						■			
5) Prepares Monthly Trade Agreement Report							■		
6) Reconciles Monthly Trade Agreement Report to accounting records				■					

FIGURE 13.2 *Sample Segregation of Duties*
Source: *BCFM Trade and Barter Guideline*—revised 2007.

Some companies use an internal form known as an advertising trade agreement to record the proposed terms and conditions, including fair market value (Figure 13.3). The completed advertising trade agreement is submitted for review and approval by the sales manager, credit manager, business manager/controller, general manager, or divisional manager, depending on policy and value of the proposal. Some organizations require corporate approvals in cases in which the trade is for significant dollar amounts or for personal consumption. Verbal approvals are insufficient.

After approval, the agreement is forwarded to the appropriate person in the Business or Finance Department (typically a market controller or business manager), who will review the agreement to assure that it is properly completed with appropriate signature(s), contains all the required information, and that the estimated value of the goods/services to be received is based on fair market value.

Station ____________________

(All information must be filled in before submitting this form to Sales Manager for approval.)

Date ____________________ Salesperson ____________________

Advertiser ____________________ Agency ____________________

Address ____________________ Address ____________________

____________________ ____________________

Contact ____________________ Phone Number ____________________

Credit Approved by Credit Manager ____________________

Specific Reason for Trade ____________________

____

____

____

Complete Description of what is to be received, and how and when it is to be used

____

____

Estimated Fair Market Value of what is to be received $ ____________________

Proposed Terms: Dollar value to be received $ ____________________

Dollar amount of airtime given $ ____________________

Trade Ratio ____________

Start Date of Agreement ____________

End Date of Agreement ____________

Remarks ____________________

____

____

Approved by Sales Manager ____________________ Date ____________

Approved by General Manager ____________________ Date ____________

Approved by Division President* ____________________ Date ____________

Approved by Station Controller ____________________ Date ____________

*Verbal Approval Obtained by ____________________ Date ____________

Trade Contract Number Assigned ____________________

FIGURE 13.3 *Sample Advertising Trade Agreement*
Source: *BCFM Trade and Barter Guideline*—revised 2007.

Once the trade agreement is approved by the required station/system/network or corporate parties, it should be finalized and a formal agreement executed. This can be done by utilizing the same trade agreement and obtaining client signatures or by using a trade contract similar to the one shown in Figure 13.4.

After execution, the contract is recorded in the general ledger by the Business Department, the airtime can be scheduled, and the receipt of merchandise and service may begin. Airtime spots must be booked and logged on the official station log.

To: ______________________________ Date: ______________________________

This letter is intended to confirm certain information with respect to a barter agreement which exists between the parties named herein. Unless I am contacted immediately, the following terms will be considered to be in accordance with your understanding of the barter arrangement.

Time on [Station] will be exchanged for [Merchandise/Services] of $_______________ in value to be supplied by [Supplier] under and subject to the following terms:

1. [Supplier] has agreed to order, and [Station] has agreed to accept, announcements of an aggregate value of $ _______________ during the term stated in (3) below. The advertising will be used for _______________. Such announcements will be valued at rates prevailing at the time such advertising is placed. All commercial material for the announcements shall be furnished by [Supplier] and [Supplier] shall be responsible for all production, duplication and integration costs associated with this material.

2. [Supplier] may order said time on a spot announcement basis only. All announcement schedules are subject to:
   a. Availability at the time of offering; and,
   b. Preemption by _______________.

3. The term of the barter agreement will be from _________ through _________. In the event that [Supplier] fails to order all or any portion of the advertising time granted hereunder within the time specified, carryover of such airtime will be at the discretion of [Station]. Generally, a carryover will only be granted if [Station] is unable to deliver said time due to lack of availability or technical difficulties.

4. In exchange, and in full consideration of the rights granted to [Supplier], [Supplier] agrees to furnish [Station] with the following merchandise/service(s):

   ____

   ____

   This merchandise/service(s) will be ordered and delivered as follows:

   ____

   ____

5. If merchandise is involved, [Supplier] hereby warrants that the merchandise delivered by [Supplier] is new merchandise in working condition.

6. [Station] and [Supplier] agree to furnish proof of performance in the form of memo billings, merchandise invoices, or service invoices on an as-provided basis.

7. All announcements placed hereunder will be subject to the terms and conditions set forth in the standard [Station] Sales Contract, in effect at the time of placement.

FIGURE 13.4 *Sample Trade Contract*
Source: *BCFM Trade and Barter Guideline*—revised 2007.

8. In no event shall the contractual obligation of this barter agreement be assigned, transferred, or offered for resale by [Supplier] to any third party, agency, or time broker without the express written approval of [Station], nor should the barter be used in any way other than in accordance with the terms of this agreement.

9. All advertisements, messages, products, or services are subject to the prior approval of [Station], and must comply with all of [Station] rules, policies, standards and practices, with all Rules and Regulations of the Federal Communications Commission and any applicable federal, state or local laws.

10. Sales and use taxes on the merchandise/service(s) supplied pursuant to this arrangement shall be the sole responsibility of [Supplier]. No agency or sales commissions on this arrangement will be payable by either party.

If the foregoing meets with your approval, kindly so indicate by signing the original and duplicate of this letter in the space so provided below and return same to us.

Yours truly,

____
Vice President and General Manager

____
[Station]

Accepted: ____________________________

By: ____________________________

Title: ____________________________

Date: ____________________________

FIGURE 13.4 (*Continued*)

Invoices should be sent to the trade client as evidence of airing. Trade revenue may be recorded in the same manner as cash expenses, or segregated into trade expense accounts.

As merchandise is received, it should be recorded and maintained in a secure area. All usage should be documented on a properly approved form (see Figure 13.5).

Trade merchandise used for giveaways is still considered income to the recipient, and is subject to federal and state tax regulations. For all giveaways, the station/system/network must obtain full name, address, and Social Security number of the recipient. The best way to do this is to require recipients to complete and sign a Form W-9 either before or upon receipt of the giveaway. The information will be used to satisfy all filing requirements governed by the Internal Revenue Service (i.e., Form 1099).

If personal use of trade is approved by station management, the user must have fair market value of the trade merchandise or service used reported to him or her and to the IRS as ordinary income for W-2 purposes. Federal and state regulations require proper withholding based on this fair market value.

Station ______________________

Date ________ **Trade Contract #**________________

Trade Client/Agency______________________________

Please check appropriate application:

☐ Receipt
☐ Distribution
☐ Shipped Direct from Supplier to Recipient
☐ Use of Services

		Trade Value			
Quantity	Specific Description	Each	Total	Percent	FMV

If Goods / Services are Used or Distributed:

Name of Recipient / User ______________________________
(Company/Individual/Position)

Address ______________________________

Reason ______________________________

Value of Related Cash Sales ________________ Sales Contract # ________

Goods/Services as a Percent of Sales Order ______ %

Percent Authorized ______% Percent Given to Date ________%

Required Documentation Attached:

☐ Purchase Order
☐ Receiving Report
☐ Shipping Document
☐ Other ______________________
☐ Acknowledgement by Recipient
☐ Delivery Receipt from Supplier
☐ Invoice/Statement

Release to Employee for Distribution to Customer:

I accept responsibility to obtain documentation (acknowledgement) of the receipt of the above merchandise upon delivery.

____
(Employee Signature)

Preparer's Signature ______________________ Date ________

Approved by General Manager ______________________ Date ________

Approved by Controller ______________________ Date ________

FIGURE 13.5 *Sample Requisition/Trade Usage Form*
Source: *BCFM Trade and Barter Guideline*—revised 2007.

A trade report that reviews the status of all trade agreements should be issued monthly (see Figure 13.1).

It is advised that a file be prepared and maintained for each trade contract. The file should include:

1. The original approved trade agreement.
2. Any contract correspondence, additions, or amendments.
3. All backup documentation necessary to support the completion of the agreement, including affidavits of time.
4. Trade-usage forms.
5. Invoices verifying receipt and use of the trade goods and/or services.

In Conclusion

Through proper internal controls, the trade or barter asset and liability accounts can be properly maintained. The objective of the trade or barter transaction, contemplated at the inception of the agreement, to improve the media company's operating results will then have been achieved.

14 Music Licensing and Syndication Fees

Mary M. Collins[1]

You work for a local all-news station that's getting crushed in the ratings with its late night programming. Looking at the competition, you realize that none of the stations in the market have any music after midnight. So you decide to show your boss that you have initiative, and you program a weekly one-hour music special that runs at 2:00 a.m. Relatively speaking, your new program is getting killer numbers, and you begin to plan what you are going to do to celebrate your raise and promotion. But you failed to consult the business manager before you put the program on the air—and now you find out that this one-hour weekly obligates your operation to a different music license. Instead of the raise, you barely escape being fired, and are advised to learn a little more about music licensing before you "take initiative" again. You agree, and vow to read what Mary M. Collins has to say on this complicated subject.

Introduction

Music is integral to media; it brands radio and television stations, networks, television programs, and news. This chapter discusses the payments for the use of this music. It is not a legal treatise on music copyrights. Instead, it is intended to raise the reader's awareness of copyright requirements underlying music payments and how these payments are administered. Although specific questions should be directed to an attorney who specializes in music copyright law, the Radio Music License Committee (RMLC), the Television Music License Committee (TMLC), and the National

1. The author would like to acknowledge Keith Meehan, Executive Director, Radio Music License Committee; Willard Hoyt, Executive Director, Television Music License Committee; Andy Holdgate, President, Holdgate Public Relations; and David Oxenford and Robert Driscoll, both of Davis Wright Tremaine, for their assistance with this chapter. Any mistakes are the responsibility of the author and the Broadcast Cable Financial Management Association.

Cable & Telecommunications Association (NCTA) can be helpful in providing some general guidance concerning the industry's agreements with ASCAP (the American Society of Composers, Authors and Publishers), BMI (Broadcast Music, Inc.), and SESAC.

U.S. music license payments have their roots in U.S. copyright law. Simply stated, copyright law gives the creator of music the ability to control the use of his or her works, and thus to realize compensation for this usage. These *intellectual property rights* are similar to those granted to patent holders. Just as an inventor may transfer rights to a patent, a songwriter or musical artist may sell, assign, or license any of his or her intellectual property rights.

In the early 1900s, U.S. copyright laws were new and often ignored when it came to music. Songwriters and publishers in this country relied on income from the sale of sheet music. To change this, a group of about 100 members of the music industry, including a composer and conductor named Victor Herbert, formed the not-for-profit American Society of Composers, Authors and Publishers (ASCAP) in 1914 to enforce their rights under the U.S. Copyright Act of 1909. Soon afterward, Mr. Herbert sued Shanley's Restaurant for performing rights. In a unanimous decision written by Justice Oliver Wendell Homes, Jr., the U.S. Supreme Court established the precedent for allowing songwriters to protect their performing rights in the music they created. Interestingly, ASCAP was not the first organization of its type. Decades earlier, in 1851 a similar lawsuit by the French composer Ernest Bourget against a café that performed his popular work *Les Ambassadeurs* was decided in his favor and led to the creation of the Société des Auteurs, Compositeurs et Éditeurs de Musique (SACEM)—the first performing rights society in the world.

From the beginning, there has been an uneasy relationship between the electronic media and performing rights organizations (PROs) such as ASCAP. In 1940, radio broadcasters formed a new not-for-profit society, Broadcast Music, Inc. (BMI), as a competitor to ASCAP because they believed ASCAP was significantly overcharging for performance rights fees and was severely limiting the composers who could join that organization. Prior to BMI's founding, composers in genres including country, blues, rhythm and blues (R&B), and other staples of today's popular music were not represented by ASCAP. With no central source for licensing, many radio stations opted not to play this music. SESAC, founded in 1930 as the Society of European Stage Authors and Composers, is the smallest of the PROs operating in the United States, and the only one that is a privately owned, for-profit enterprise. Originally an agent for European publishers whose works were not being represented in the United States, the society changed its name to SESAC, Inc. in 1940. (Today, ASCAP, BMI, and SESAC all have open-membership policies that allow most published composers from every genre to join their organizations.)

Overview of Music License and Syndication Rights

In general, broadcast and cable users of music must pay two separate fees: one for copying or using music as part of their commercial or program productions, and the second for transmitting their programming to the public. Thus, a media company that produces a program or commercial must pay for the right not only to use the music in its production activities, but also for the performance of the music as part of its programming. Using music in radio commercials requires permission to copy the music; the use of music in television or cable programs or commercials requires permission to synchronize the music. (It should be noted that, as of this writing, a group including representatives from record companies, recording artists, and bands is campaigning for a third "royalty" payment from radio stations. The industry refers to this as a new "tax" on their business.)

Performance Rights

The public performance right, or simply "performing right," is the composer's right to be compensated for the public performance of his or her musical composition. Whether the music is distributed in a broadcast, via cable television, or via a digital-communication platform, the distributor must pay for the right to play the composition. This performing right should not be confused with the *audio digital performance* right designed to award the recording artist, not the composer, of the music. Traditionally, composers transfer their copyright rights to publishers responsible for authorizing many uses of the music, and retain the right to be paid directly for half of the performing rights. Composers and publishers generally assign the right to collect their performing rights to one of three PROs: ASCAP, BMI, or SESAC. These performing rights organizations charge stations and cable systems a license fee for the broadcast, transmission, or other communication to the public. These industry-wide licenses have always allowed licensees to use any music within a respective PRO's repertoire under a so-called blanket or per-program license. These licenses differ only in the method by which the fees are charged for the clearance of all of the music in the repertoire (see below).

Production Rights

Licensing the right to broadcast or cablecast music does not allow a station or cable system the right to use that music to record a commercial, jingle, logo, or locally produced program. That use requires a second copyright license, which has to be obtained from the copyright owners or their agents.

One particularly troublesome area is the use of music in commercials. A popular recorded song cannot be used in a recorded local commercial without obtaining the rights from the copyright holder in the composition and from the record label that owns the rights in the master recording. Nor can the song be rerecorded by a local group and used in a recorded local commercial without obtaining the rights from the owner of the copyright in the composition.

Even what might be thought of as a "parody" of a song—sung to the tune of a popular song, but promoting the products of a local merchant—can be a problem. Although parodies are protected by copyright law, the parody must meet several tests before it is protected, including the fact that it is making fun of the copyrighted work itself. A commercial for a product using the tune of a popular song, even if it is funny, may not be protected if it is not making fun of the underlying song, but is instead simply being used to promote a product.

Radio

The use of a prerecorded song as part of a radio commercial produced by the station requires the producer to obtain a license for the "sound recording," which generally is licensed by a record company, as well as a separate license for the use of the musical work, the basis for the recording. In order to use a musical composition and not the specific recording of a song, the radio station must obtain the right to copy or record the song. Neither of these licenses is obtained through one of the PROs. Instead, they must be obtained, in the case of the musical work, from the composer or publisher—and, in the case of the sound recording, from the record company. It should be remembered that composers, publishers, or record companies are not required to issue a license, and may deny use or demand hefty fees in order to permit the use in a given circumstance. This is especially true when a composer or recording artist wants to protect the integrity of his or her song or recording, and thus may make it difficult to get permission for use in a local commercial spot production. It is a good practice to make initial contacts with the composer, publisher, or record company before committing to use a given song or recording in a commercial.

Television and Cable

In television and in cable, using music in locally produced programming and in commercials requires a synchronization right—that is, the right to use the music in synchronization with video. This right is normally controlled by the composer or the publisher, and is not licensed through the three PROs. Again, making the necessary contacts to get the right to use the music before making plans to include it in a commercial or a local program can save a lot of time and money.

Network Licenses

It should be noted here that the producer of programming, recordings, commercials, logos, and promos is generally responsible for clearing the rights to use the recording or synchronization, but the local station or operator is generally responsible for the performing rights. This is not always true where the programming is produced by a broadcast or cable network. The major broadcast and cable networks historically have licensed directly with ASCAP, BMI, and SESAC to cover the performing rights of the music in their network programming "through to the viewer." Because these networks also produce the programming, or are responsible for producing it, a local station or operator doesn't have to pay for performing rights for this programming. These network performing rights do not cover the music in commercials sold by a local station, whether those commercials are local or national spots. Therefore, if a local station or cable system also runs commercials inside the programming, it can be responsible for the performance of music within these commercials. Thus, a station has to be careful to make certain that all of the programming and the commercials, promotions, and logos are cleared for performing rights if these rights are not otherwise licensed.

In summary, public performance rights are generally granted by one of the three performing rights organizations: ASCAP, BMI, or SESAC. Musical rights required to produce local programming, commercials, promos, or other material broadcast or cablecast are in addition to performance rights, and different copyright clearances are required for different media and/or different uses. Music copyright law can be very complicated. It cannot be stressed enough that an attorney familiar with music copyright law is the best source to answer questions about the various copyright requirements for a specific use in a production. Moreover, the safest practice is to consult the attorney before committing to the use of the music.

Online and Other Rights

When webcasts—including advertisements and on-air promotions—contain licensed compositions or recordings, there are additional rights to be considered. These rights are continuing to evolve as new distribution methods are created. The following will provide a general overview of considerations. As is the case with other music rights, it is advisable to confirm the rights before using the composition.

Radio

Radio stations need to address three general guidelines for repurposing their over-the-air content for online use:

- Observe certain standards set by the Digital Millennium Copyright Act of 1998 (DMCA). These standards address such issues as registering with the U.S. Copyright Office for a compulsory license, record-keeping requirements, restrictions affecting advance promotions of a specific song, and limitations on the number of music titles for a particular artist or album that can be performed consecutively or within a specific time period.
- Pay the announced rates established by the PROs for the use of musical compositions on the Internet. ASCAP, BMI, and SESAC have created licenses for Internet use of licensed music. If the station is simply retransmitting its over-the-air signal, the ASCAP and BMI broadcast licenses cover Internet use, but SESAC requires a new license. All three societies require a new license when the radio station is creating a new programming stream for the Internet. Although each agreement is different, all of them involve a minimum fee and royalties based upon revenues or web site usage.
- Pay royalties to SoundExchange for the compulsory license for the use of sound recordings. SoundExchange, a cooperative of copyright owners of sound recordings (with significant participation from the major record labels), oversees a new copyright-payment requirement affecting digital music. The DMCA requires that royalties be paid to performing artists for all digital uses of music except those in an over-the-air digital transmission by a broadcast station, such as iBiquity Digital Corporation's digital-radio systems. Half of the fees paid to SoundExchange goes to the performers featured on a recording; the other half goes to the copyright owner of the performance, which is usually a record company.

Over-the-Air Television and Cable Television

In addition to the synchronization rights for reproducing a musical composition and the master use license required for using a recording of a song when that song is used in conjunction with a recorded video production, cable and television companies will likely need to get additional rights from copyright holders before video productions containing music can be used on a web site.

It is important to make sure that the rights to use music in a video piece are broad enough to cover the repurposing of that content for online or other uses. There have been a number of issues raised in many contexts, not just music, in which rights to a copyrighted work have been obtained for one medium, and one medium only, and do not extend to the use of those works on the Internet. This has particularly been true in the case of older productions that were done before video in the Internet was even a gleam in producers' eyes. The rights granted for the use of music in a video production may need to be extended before that production can be used on a web site or other medium.

User-generated content provides another area of potential concern. Web sites, including those for television and cable companies, may give viewers the opportunity to post their own video productions. While copyright law provides some protection to parties that host an online bulletin board-type of service from being subject to copyright liability for infringing uses posted by third parties, as of this writing, the law is still being tested in cases where users post their own video productions.

Moreover, the law imposes duties on the bulletin board provider before he or she can claim these protections. The requirements include that the provider act promptly to remove any infringing material once that provider is put on notice that the infringing material exists. In addition, the provider cannot encourage or promote the infringement.

With that in mind, terms of use for sites that allow viewers to post their own videos should be carefully drafted to warn users that they should not use copyrighted material without permission. Companies should also be careful about contests that might be seen as encouraging copyright infringement (e.g., a "make your own music video" contest, unless the company has the rights to use the song or songs that the videos will feature).

Putting content online, whether it is a simulcast of a radio station's over-the-air broadcast or specific video programs from a television or cable company, also exposes the content to a far wider audience, often making it available to viewers who might notice a copyright infringement that local audiences would overlook. So companies should be extra careful about all uses of music that might show up online.

PROs' Responsibilities to the Broadcast and Cable Industries

Both ASCAP and BMI currently operate under antitrust consent decrees. For broadcast and cable, the most important provisions of these decrees are:

- The requirement that ASCAP and BMI grant interim licenses while negotiating final rates for performing rights.
- The ability of either party to appeal to a federal court to decide rates and other provisions if licensees and ASCAP or BMI cannot agree on license fees and/or other provisions of a final license between the PROs and any respective group of licensees.
- Neither ASCAP nor BMI may obtain exclusive rights from composers, thus allowing composers to sign individual direct licenses with licensees in place of the licenses offered by their respective PRO.
- Both ASCAP and BMI must offer per-program licenses, the fees for which are based on the use of ASCAP or BMI music in particular programs.
- A requirement that these PROs must offer comparable licenses to stations that are "similarly situated," at reasonable license fees, which means that radio and television stations that meet the notice requirements of the decrees are protected against copyright-infringement suits by either ASCAP or BMI.

SESAC, the third PRO operating in the United States, and the only privately owned, for-profit society, is not subject to a consent degree as of this writing.

Performing Rights Licensing and Fees

Overview

Negotiations with the PROs can be conducted by a committee set up to represent a specific group, by a parent company, or by the individual station or system. In the 1930s, the National Association of Broadcasters (NAB) organized a group of radio representatives to negotiate industry-wide fees. This ultimately led to the creation of the current Radio Music License Committee (RMLC), which negotiates with ASCAP and BMI—but not SESAC—on behalf of the majority of U.S. commercial radio stations. The Television Music License Committee (TVMLC) is also an NAB offspring.

This committee negotiates with ASCAP, BMI, and SESAC on behalf of most commercial television stations.[2] These two committees are funded by participating stations or station groups—the radio committee receives court-ordered assessments, and the television committee relies on voluntary contributions for its funding.[3]

In the cable industry, the National Cable & Telecommunications Association (NCTA) represents cable operators in negotiations with ASCAP, BMI, and SESAC. And, as indicated below, the majority of cable programming companies have negotiated their own licenses with the PROs.

The following information is not intended to be an exhaustive discussion of fee-calculation methodology, but rather, to provide the reader with an overview of the process for determining fees. For specific information, contact the group, department, or individual responsible for negotiating the agreement in question.

Radio

Blanket License Fee

This is the most commonly used license in radio (it is also the most common license for television and cable). ASCAP, BMI, and SESAC all offer radio stations a blanket license fee under which any of the music in the individual society's repertory can be played in exchange for license fees. The RMLC administers the fees for ASCAP and BMI, using a complicated formula designed to guarantee both societies a specific total amount of fee revenue based upon a schedule agreed to by both the individual PRO and the RMLC.

SESAC's blanket fee is negotiated directly by the radio station or its parent company. In general, the SESAC blanket rate is determined by the individual station's market—typically the population of its MSA (metro service area) or county—and its high one-minute spot rate.

Radio Per-Program Fees

Radio stations that do not perform much music—that is, stations with a news, talk, sports, business, or other format that broadcast feature music in less than

---

2. The TVMLC negotiates only on behalf of full-power commercial television stations; it does not represent low-power television stations (LPTV) or public broadcasters.

3. Another group, the National Religious Broadcasters Music License Committee (NRBMLC), represents approximately 500 radio stations; the terms of the license negotiated by this committee are different from those negotiated by the RMLC. Music license agreements for public radio stations are also outside the responsibility of the RMLC.

approximately 30 percent of their day-part programming—may opt for a per-program license under the RMLC's agreements with ASCAP and BMI. The fees paid by these stations are calculated annually by the RMLC as part of the overall industry license fee owed to ASCAP or BMI.

SESAC offers stations that feature programs consisting primarily of news, talk, sports, or other content devoid of feature musical presentations the opportunity to qualify for an all-talk amendment that reduces the license fee the station would otherwise owe.

Television

Blanket License Fee

As in radio, a blanket license fee allows the television broadcaster the right to publicly perform any of the music written by a composer who is a member of one of the PROs.

Once an industry-wide blanket license fee is determined for a particular license year, the TMLC determines, with approval from the PROs, how the industry fee is allocated to individual stations. Historically, the ASCAP and BMI allocations have been based on market size and station audience levels, with a prime-time audience credit to broadcast network affiliate stations, whose network programming is separately licensed by their respective networks. Like industry-fee negotiations, these allocation methodologies are partly a result of historical compromises within the television industry.

The SESAC blanket fee allocation, as of this writing, relies primarily on individual station music use as the basis for determining a station's share of the blanket fee. The committee's formula includes a "recurring use" performance unit to determine the value of the television programs that include at least one minute of SESAC music in 75 percent of their episodes. These program values are then used to allocate fees among stations by using audience levels and minutes of music in these programs. The remaining "occasional use" portion (20 to 30 percent) of the fee is allocated employing the same method used for ASCAP and BMI.

TV Per-Program License

Although the television per-program license, like the blanket license, grants the same full clearance of all of the music in a PRO's repertory, the per-program fee is calculated differently. Under this license, a station has to pay fees only for programs that contain a specific PRO affiliate's or member's music, although the rate for each program is higher than the blanket fee rate. The per-program license is used mostly by broadcast network affiliate stations—partly because their networks separately license their network programming, and partly because they produce a lot of local news as part of their nonnetwork programming.

The reason that producing local programming is so important in terms of qualifying for per-program savings is that, unlike the music in network or syndicated programming, local stations can control the music used in these programs. A station may make a business decision to avoid paying a music license fee for a particular program. In this case, the station must make certain that there is absolutely no music composed by a member of the respective society in that program unless that music is "otherwise licensed." Even one second of music in a program will make the entire program subject to a fee. This means that if a program contains one second of ASCAP music and 15 minutes of BMI music, a station pays the full fee related to that program to both ASCAP and BMI, rather than a pro rata share based on each society's portion of music in the program.

In general, the calculations of all three of the PROs' per-program license fees involve multiplying the station's blanket license by a multiplier to set a new per-program base fee. Once that new higher base fee is set, revenues applicable to local programs that contain the society's music are determined as a percentage of the local station's total revenue applicable to all of that station's local programs. That percentage is then multiplied by the per-program base fee to determine the net per-program fee. In addition, a portion of the per-program fee is apportioned to "incidental" uses of music. Additional information about per-program licenses is available from the TVMLC.

Direct Licensing/Ownership of Copyrights for Local Television Programming

Stations may opt to license theme and background music in local news programming directly with a composer, and thus benefit from savings under the per-program license. In this case, stations should be careful to make certain that both synchronization and performance rights are cleared at the time of the negotiations. A lawyer familiar with music copyrights should be consulted to make certain that the license is properly worded.

Stations may also want to consider the possibility of *owning* the copyright to music, rather than merely licensing that music. Licenses are typically related to a specific use on a specific communications platform, rather than the right to use the music on any platform communicated anywhere in the world. Given the dramatic changes in the television and cable businesses and the use of digital transmission from various platforms, it is difficult to determine what clearances will be required in the future.

Because direct licenses normally give exclusive rights to the music within the station's DMA (designated market area), composers can be compensated for the same music in several different markets. That means they are typically more willing to license the music than to sell the copyright. However, it is possible under either a license or a sale to provide for unlimited use and performance by the station, while still providing an opportunity for the composer to use the same music in other markets. Again, a lawyer experienced in music copyrights should be consulted to provide guidance as to whether licensing or purchasing a copyright would be preferable.

Source Licensing

Source licenses are agreements with program owners to clear performance rights in a particular program. They can benefit users in the same way that direct licenses do in terms of the per-program license. These source licenses are generally made between program syndicators and an agent working with local television stations. Because the agent normally determines the fees and the provisions, stations generally need determine only whether or not the portion of the source license charged to the station is more or less than the per-program savings for the station.

Future Television Licenses

Television stations are converting from a single analog channel to a digital transmission that provides multiple digital-broadcast signals. Thus, the programming on these new signals requires performance licensing. The current industry blanket licenses with ASCAP and BMI provide coverage for these digital signals and for some programming on the Internet. The current SESAC industry license provides coverage only for one signal, and only if that signal simulcasts the analog signal. As of this writing, the TVMLC has reached an agreement with SESAC not to sue in conjunction with SESAC music transmitted via Digital Media on any station represented by TVMLC provided that the stations make the appropriate payments to SESAC (for more information go to www.televisionmusic.com).

The committee has also applied on behalf of stations for a video on demand (VOD) license with ASCAP and BMI, and has requested a VOD license from SESAC. The application for the ASCAP and BMI licenses requests protection from copyright infringement for local stations; final terms were not available as of this writing.

Cable

Cable Programmers

As in broadcasting, there are both programmers and distributors in the cable industry. The programmers are typically known as cable networks; they include such channels as CNN, ESPN, Discovery, MTV, and any of the other many, many channels available on a cable system. ASCAP, BMI, and SESAC have all negotiated blanket through-to-the-viewer licenses with the majority of premium and basic cable networks, as well as with many VOD program providers. To assure that similarly situated networks pay similar fees, the PROs' license fees for cable networks are generally based on each network's subscriber base, net advertising and subscriber revenues, Nielsen ratings, and type of programming provided.

Cable Operators

Cable operators operate the cable system that delivers programming to subscribers' homes. The NCTA has negotiated licenses with ASCAP, BMI, and SESAC on behalf of NCTA members. The licenses authorize the performance of music in locally originated programming, including leased access and PEG (public, educational, and government) channels. Under the NCTA-negotiated agreements, cable operators pay a per-subscriber license fee to the PROs.

AFTRA

The American Federation of Television and Radio Artists (AFTRA) is a national labor union representing performers, journalists, and other artists working in the entertainment and news media. Although they are not strictly either music license or syndication fees, it is important to be aware of AFTRA's contract terms because they can add significantly to program costs. AFTRA's contract stipulates minimum compensation as well as terms including the number of weeks during which the spot can be run. Current AFTRA agreements cover: "Radio Recorded Commercials," "Commercials Made Intentionally for the Internet," "Internet Use of Commercials Made for Initial Use on Broadcast Radio," and a variety of video applications.

In Conclusion

Music has been, and will continue to be, an important part of electronic media. U.S. copyright laws guarantee the right of the creator of the music, as well as that of the entity that performed the composition, to be compensated for their work. Performance rights are generally granted by one of the three performing rights organizations, ASCAP, BMI, or SESAC. Musical rights required to produce local programming, commercials, promos, or other material broadcast or cablecast are in addition to performance rights, and different copyright clearances are required for different media and for different uses.

With the continuing evolution of distribution technologies, companies must take special care to ensure that they have secured the necessary rights before using music on the radio, on television, in cable programming, on the Internet, or by means of any of the myriad of other technologies available to today's consumers. Whenever there is the slightest doubt about the license required, contact an attorney familiar with music copyright law. To do otherwise means risking additional fees or even being the subject of a court decision quoted in future articles, chapters, or books about music licensing.

15 Taxation in Brief

J. Michael Hines, Esq., and Geoffrey J. Christian[1]

Nothing is more certain than taxation. Well, maybe one thing. And anyone in business has to deal with a wide variety of taxes. Sales and use taxes, real estate taxes, payroll taxes, income taxes, and the rest are the stuff that governments run on. Businesses must develop wise and adroit ways of paying just the right amount of taxes, no more and no less than what is required. In this chapter, Mike Hines—a Washington, D.C., tax attorney and member of Dow Lohnes PLLC—and Geoff Christian—a CPA and tax accountant and member of Dow Lohnes Price Tax Consulting Group LLC, located in Greenville, S.C.—guide us through the convoluted paths to the right and perfect tax payments.

Introduction

Large investments are required in order to own and effectively operate radio, cable, and television properties. As a result, a business manager or financial manager must take an active role in the daily management of the company. The financial manager knows that better planning and prudent, aggressive decision making can save significant amounts of money. Taxes are one such area.

It's not unusual to think that most taxes are fixed, at least to a degree—so what can be done to change things? Even if a manager does not prepare a single return, he or she will make or influence decisions that affect the outcome of the taxes that the company owes. In fact, a company may spend a higher percentage of its gross revenue on taxes than on any other single item of expense, with the exception of personnel and program costs. It is management's responsibility to keep that percentage as low as possible, within the bounds of applicable tax rules.

Obviously, a few pages in this book will hardly make a dent in the tax laws. In an attempt to be as efficient as possible, this chapter will cover these ideas on a topical basis. Remember, these are general concepts. Actual tax-planning ideas are best implemented after management evaluates both the potential long-term benefits and how the idea applies to the specific company.

---

1. The authors are indebted to Ralph Bender, CPA, whose original version of this chapter was published in the first edition of this handbook.

Property Taxes

Property taxes are a local issue, with rules varying from one area to another. However, as federal aid to state and municipal governments shrinks, local governments increasingly target property taxes as a rich source for revenue increases. Property taxes are traditionally categorized into three areas: real property, personal property, and inventory. Real property (land and buildings) tax assessments are generally based on construction cost, purchase price, or market valuation, and offer only limited room for negotiation. Special features such as the studios of a broadcast facility, if taxed on a fair market value, may reduce the feature's value to any buyer except to another broadcaster. On the other hand, personal property taxes offer many opportunities for savings. Most personal property taxing authorities offer suggested guidelines for useful asset lives. Shifting assets from a category with a long useful life to a category with a shorter useful life can result in reduced taxes. Companies may benefit from careful evaluation of the category or class to which assets are assigned. Often the local assessor has some flexibility. Maintaining good records will help in ending tax payments on assets that are no longer being utilized. When allowed by local ordinance, company records will assist in removing assets that are sold, otherwise disposed of, or no longer in use. Inventory taxes, though normally not a significant cost item except perhaps to cable operators, are similar to personal property taxes. Normally based on an average year-end cost, inventory taxes are reduced by keeping year-end inventory to a minimum. Some assessors will also allow reductions based on the cost of waste incurred in inventory use.

As discussed in the next section, Depreciation, expensing assets of relatively small amounts is a common practice. Typically, property tax returns (or renditions, as they are sometimes called) are based on fixed asset records. If low-dollar assets are expensed rather than capitalized, the items are not captured in the fixed asset records, and most likely will be left off the property tax return unless a procedure is in place to add those items to the return. The auditors may review common expense categories with a view toward adding back expensed assets.

Media companies may hire an outside consulting or accounting firm to assist in classifying assets, evaluating idle assets and obsolete inventory, and challenging assessments. The involvement of these firms can range from simply providing assistance to assuming the property tax function completely. Fee arrangements vary depending upon the level of involvement of the outside firm, so it is important to take the potential savings generated by improved classification and reporting and compare those savings to the fees for the service.

Depreciation

The regulations governing depreciation and amortization have been the subject of frequent legislative changes. Whereas business seeks to reduce the acceptable length of time over which an asset may be depreciated, preferring to use more-accelerated, or front-loaded, methods, the government takes in tax revenue earlier if the depreciation period is lengthened and straight-line or less-accelerated methods are used. As a consequence of this tug-of-war, depreciation rules have become a complex maze. Like any maze, though, one who knows the correct path can spend the least time and achieve the best available result.

Although really a component of the income tax computation, tangible and intangible asset depreciation (usually called "amortization" in the case of intangible assets) is significant enough in its impact that it will be discussed here on its own. The three primary Internal Revenue Code sections governing depreciation are Code Sections 167, 168, and 197. Depreciation periods and methods can vary greatly by the type of asset and its intended use. Because this is a complex area, companies should review the preliminary determinations with a tax adviser before establishing the depreciation (or "cost recovery") categories of assets.

For assets placed in service before 1981, there were several allowable methods of depreciation—generally the methods allowed for accounting purposes. Tangible assets placed in service from 1981 through 1986 are depreciated using ACRS (Accelerated Cost Recovery System). Tangible assets placed in service after 1986 are depreciated using MACRS (Modified Accelerated Cost Recovery System). ACRS and MACRS are both methods that assign specific depreciable lives and accelerated-recovery methods to classes of assets. The amount of depreciation allowable each year is computed using percentages from IRS-provided tables.

Because of the use of accelerated methods instead of straight-line depreciation for most tangible personal property (such as equipment and vehicles), the difference between the accelerated amounts and the straight-line amounts must be computed and reported as an income tax preference item for purposes of the Alternative Minimum Tax (AMT). If presented in accordance with Generally Accepted Accounting Principles (GAAP), the financial statements will likely require another depreciation computation. With the very real prospect of a number of depreciation calculations per asset, the computerization of asset depreciation records will almost certainly be required in order to manage the volume of data.

One way media companies look to reduce the volume of calculations is to expense small-asset purchases instead of capitalizing them. By immediately expensing small items, the company gains immediate deductibility and reduces the record keeping associated with depreciation. It is important to establish a capitalization policy that provides this flexibility. This policy should be in writing, and should state that all capital assets purchased with a cost less than a specified amount per item must be charged to expense when acquired, without exception. The IRS has generally approved "reasonable" capitalization policies. It would be prudent to check with a tax adviser before implementing a minimum-capitalization policy.

In addition, Code Section 179 allows a taxpayer to expense in the year of acquisition certain amounts of the qualifying personal property purchased and placed in service that would otherwise be capitalized. However, there are restrictions under Section 179 that could limit the amount that can be expensed, and this provision is subject to legislative change.

Another area of potential tax relief deals with real property component depreciation. The cost of identifiable tangible personal property components of real property—such as heating, ventilation, and air-conditioning systems; and elevators and escalators—can be separated from the real estate and depreciated over a shorter life. With real property depreciated over a relatively long useful life of 39 years, this is an important technique to accelerate the depreciation of capitalized items.

Another area that may apply is the depreciation of luxury vehicles. The IRS places limits on the amount of depreciation allowed on luxury cars. The dollar limits defining a luxury car and the dollar limits on the depreciation allowed change annually. This information can be obtained from the IRS or the company's tax adviser. With these limits, it could take a significant number of years to fully depreciate a luxury car. In order to recapture some of the reduced depreciation, the business may consider selling luxury autos instead of trading them in. Under Section 1245 of the Code, the loss on the sale of such a business asset is considered an ordinary loss, provided total losses exceed gains from the disposition of such property.

Leasing vehicles has become a popular alternative to purchase for both tax and financial outlay reasons, and should be regularly considered to determine if it might be an advantage.

In 1993, a new section of the Internal Revenue Code was added that proved to be very beneficial to most broadcasters and cable operators. Section 197 provides that intangible assets acquired in connection with the purchase of the assets of a business are generally amortizable over 15 years on a straight-line basis. Almost all intangibles are eligible for such amortization, including FCC licenses, network affiliation agreements, program contracts, cable television franchises, going-concern value, workforce in place, customer- and market-based intangibles, patents, copyrights, trademarks, and trade names. Self-created intangibles generally are not eligible for the amortization deduction.

However, program costs may continue to be recovered under the more favorable income forecast method. This method is now detailed, with limitations, in Section 167(g).

Income Taxes

Following are a few income tax items, in addition to depreciation and amortization, that will affect most taxpaying entities on a regular basis.

Bad Debts

If a business is not using an allowance for doubtful accounts (a method of calculating bad debt losses for tax purposes), the IRS does not usually allow a write-off until all reasonable means of collection have been attempted. This suggests an additional tax incentive to aggressively pursue bad accounts. In this case, the media company must be certain to document the reasons for its inability to collect the receivable (e.g., business terminated, bankruptcy, etc.).

Charitable Contributions

Although the limit on corporate charitable contributions is 10 percent of taxable income, charitable contributions in excess of this amount may be carried forward to future returns. Accumulating excess contributions subject to carryforward may not be prudent from a tax perspective, however, because the benefit of their tax deduction will not be realized until future periods.

Entertainment and Business Expenses

With few exceptions, only 50 percent of all food and entertainment expense is deductible. When expense reports are submitted, employees must document who was entertained, where, when, and for what business purpose. If a flat entertainment and travel allowance is provided, the business must report these amounts on each employee's W-2 form.

A standard mileage rate for individuals for deducting automobile expenses is published annually by the IRS. Companies may reimburse employees using this rate without issuing a W-2 if mileage reports are submitted. Companies often compare this method with the relative advantages and disadvantages of leased or company-owned and -maintained vehicles.

Estimated Income Tax Payments

To avoid penalties for underpayment of income tax, a corporation must make quarterly payments of estimated taxes. Accurately calculating such estimates is important to cash management. Estimated payments are generally due on the 15th day of the month of the 4th, 6th, 9th, and 12th months of the company's fiscal year. If taxable income existed in the prior fiscal year, and the company is not a large corporation as defined by Code Section 6655 (a corporation that has had $1 million in taxable income in any one of the prior three tax years), the quarterly estimated payments should be at least 25 percent of the lesser of the current year tax or the prior year's tax as shown on the prior-year return.

Alternatively, quarterly estimated tax payments may be based on estimated annualized earnings, a method that may be advantageous if earnings are not level over the year. These annualized-earnings estimated payments must be for 100 percent of the tax estimated through that portion of the year. For example, for the first installment, based on the first three months of the year, pay one-quarter of the tax estimated to be due for the entire year.

Life Insurance

Except for group term life insurance benefiting employees, life insurance premiums are not deductible, and life insurance proceeds are not included in income; however, they may be considered in Alternative Minimum Tax calculations.

Controlled Groups

If a company is a member of a "controlled group of corporations," as defined in Code Section 1563, the company must make an election to apportion the surtax exemption, the Alternative Minimum Tax exemption, the environmental tax exemption, and the Code Section 6655 limitation. The apportionment is at company discretion. In addition, component members of a controlled group are treated as *one taxpayer* for the purposes of determining the tax imposed under Section 11. Each graduated income bracket is divided equally among the members unless they consent to an apportionment plan.

Sales and Use Taxes

The majority of U.S. states impose some form of sales and/or use tax on sales transactions. Traditionally, sales taxes have applied to retail sales of tangible personal property. Sales of services, including advertising services, were generally not subject to sales tax. Thus, many electronic communication operators had no duty to collect sales taxes from customers except when selling tangible property such as tapes or promotional items. Cable operators pay local or state franchise fees, and may also be subject to taxes similar to telephone or sales taxes on some or all cable subscriber charges. Similarly, on the buying side, such operators traditionally pay sales taxes on tangible property used in the business. However, the states have always had important differences among themselves, and the pressure for revenue has caused the scope of these taxes to be expanded and exemptions to be cut back.

Each year, most states have some sort of legislative activity adding, changing, or deleting transactions subject to sales tax, as well as possibly changing the tax rates imposed on transactions. This is an area that smart media companies will monitor closely. As state revenues decline, states look for ways to boost tax revenue, and sales tax is one of the targets, particularly now that Internet sales have made a dent in state sales-tax revenues. Technology has impacted sales taxes in ways never dreamed of in decades past. New technology and the new products that follow create new sales-tax issues faster than they can be resolved. Each state must determine whether and how to tax sales of both tangible and intangible property that never existed before, and so may not fit the definitions of existing law and regulations.

The Streamlined Sales Tax Project, sponsor of the Streamlined Sales and Use Tax Agreement is an example of states' efforts to respond to the rapidly changing tax environment. While fewer than half of U.S. states are members as of this writing, more states are slated to join in future years. The purpose of the agreement is to simplify sales-tax administration for both sellers and the states. Each company must have in place procedures to collect tax in each "member" state where the company may have a sales tax collection or use tax payment responsibility. Visit www.streamlinedsalestax.org for more information.

Because many states offer a variety of targeted exemptions from sales and use tax, media companies must be familiar with the exemption statutes in the state(s) where they operate or provide services. For example, Georgia provides radio stations with an exemption from sales tax for digital-broadcasting equipment purchased before the earlier of the date on which the radio station ceases analog broadcasting or November 1, 2008. Louisiana, on the other hand, allows television broadcasters an exemption for one purchase of digital-transmission equipment from each category provided in their statutes. To emphasize, all states do not tax or exempt the same services and/or purchases in the same way. The form in which charges are incurred can significantly impact the ultimate sales-tax liability from state to state. For example, if freight and/or labor costs are separately stated on an invoice, these charges may be exempt from tax in certain states. Again, it is imperative for media companies' financial managers to understand each state's application of their respective sales- and use-tax rules to their particular business operations in order to avoid overpayment.

Payroll Taxes

All businesses with employees pay Federal Insurance Contributions Act (FICA) payments, Federal Unemployment Tax Act (FUTA) payments, and state unemployment tax. The wage bases for the Social Security tax and the Medicare tax are published annually by the IRS. Company finance employees are responsible for making all payroll tax deposits on a timely basis. For large corporations (deposits in excess of $100,000), deposits must be made within 24 hours of the date the payroll is paid. Failure to make federal deposits on a timely basis can result in substantial statutory penalties that are unlikely to be waived. For corporate officers, an added incentive to make timely deposits is Code Section 6672. This section allows the IRS to assess a penalty (equal to the total not collected) against anyone who is required, and who willfully fails, to collect, account for, or pay any tax due. The IRS shows little restraint or leniency when collecting payroll related taxes and penalties.

In some payroll situations, employers attempt to avoid the burden associated with payroll taxes by classifying employees as independent contractors, who then must themselves pay the taxes applicable to their earnings. A general rule for the distinction is that the employer has the right to control or direct only the result of the work done by an independent contractor, and not the means and methods of accomplishing the result. Studies have shown that many workers are incorrectly classified, resulting in large tax losses to the government. Consequently, the government is continuing strong enforcement efforts to stem this flood of lost revenue.

Media company managers are advised to consult a tax adviser before making extensive use of independent contractors. If determined to have incorrectly handled independent contractors, the company may be held liable for all FICA and withholding taxes due for those contractors. However, if reporting requirements were followed in good faith, Code Section 3509 provides a lesser liability compared to cases where the requirements have been disregarded.

Excise Taxes

Many businesses have insurance, particularly slander and libel, through a foreign carrier. In this case, unless exempted by the provisions of an income tax treaty, a federal excise tax must be paid based on the premium. Additionally, federal excise tax is charged on local telephone charges. Federal excise tax is no longer imposed on toll (long distance) charges.

Information Returns

There are a number of information returns required to be filed annually, including Forms W-2, 1099 (for independent contractors, interest, dividends, and other miscellaneous income), and 5500 (for pension and profit-sharing plans and health-and-welfare plans). Even though these returns require no tax payments, failure to file on a timely basis can result in substantial penalties.

Accumulated Earnings Tax

If a business is fortunate enough to have a large retained earnings balance, the company should schedule and hold regular board meetings and carefully document in detail its future plans with regard to accumulated earnings. Code Section 531 imposes a 15 percent tax on excess accumulated earnings.

In Conclusion

Although a business can never avoid taxation completely, being aware of and alert to planning business operations within the limits of laws and regulations can minimize tax expense. Companies are responsible not only for federal taxes, but also for adhering to the requirements of state and local tax codes for each state in which they do business. In the post-Enron business environment, the Sarbanes-Oxley Act of 2002 has imposed stricter standards on public companies for ethical business practices by requiring detailed explanations of and procedures for sufficient internal controls (see Chapter 10). In this environment, it is imperative to closely analyze and clearly disclose the risk of tax positions taken. Although nonpublic companies are not currently required to follow Sarbanes-Oxley, the statute's standards are increasingly being voluntarily applied by businesses and their advisers, especially the accountants who audit their financial statements. Generally, the overall business environment is much less tolerant of highly aggressive tax planning. Given the complicated nature of taxes, many media companies have found that dollars spent on a professional tax adviser are a wise investment.

16 Financial and Accounting Considerations for Broadcast and Cable Acquisitions

John S. Sanders

The business you started in your garage a few years ago has been astonishingly successful! So much so that you attracted the attention of the major players in your field. One makes you an offer you can't refuse, and you've just pocketed a check for several million dollars. What to do with all that capital? You don't want to give it to Uncle Sam. You look around and think it would be nice to be in the broadcasting or cable field. John S. Sanders—a founding principal in Bond & Pecaro, Inc, a Washington, D.C.–based consulting firm specializing in providing financial, economic, and valuation services, including fair market valuations and purchase price allocation reports to media and communications companies—will walk you down that often slippery slope of acquisitions.

Introduction

In recent years, the financial and accounting aspects of completing a broadcast or cable acquisition have become increasingly complex. This complexity has been driven by two primary factors.

The first is greater economic and competitive pressure on the broadcast and cable industries, which has altered dramatically the environment for media companies. Historically, television and radio businesses were characterized by stable growth rates and profit margins, but they are now adapting to a more volatile environment defined by competition from the Internet, cable, and satellite services, and from portable listening devices, such as iPods. Slowing advertising growth and increasing cyclicality due to political advertising have also affected the industry.

The cable industry has also changed dramatically. What had originally been a business focused on providing one-way video content has evolved into a diversified telecommunications provider offering a "triple play" of video, Internet, and telephone services—or even a "quadruple play" that includes cellular phone service. The confluence of these economic factors has made the valuation process more complex.

The second factor that has influenced the acquisition process is the increasing financial scrutiny that has resulted since the implementation of the Sarbanes-Oxley Act of 2002 and related accounting standards such as Statement of Financial Accounting Standards (FAS) Nos. 141, 142, 144, and 157. (See Chapter 10.) These measures were implemented in the wake of high-profile financial scandals at Enron, Tyco, Adelphia, WorldCom, and other companies. The measures impose strict standards on how companies report the value of acquired assets and liabilities and, on an ongoing basis, how they convey information to investors about the validity of those values. Moreover, auditors are now devoting greater resources to reviewing and analyzing the financial accounting associated with an acquisition.

This is a brief overview of the financial and accounting considerations in a broadcast or cable acquisition. It is not intended to be an in-depth treatise or a step-by-step guide to accounting for a transaction. Each transaction will have its own unique characteristics; experienced buyers of media properties know that one of their best investments when going through the process is hiring someone who handles these types of transactions for a living.

The areas to be covered include determining the value of a property, due diligence considerations, purchase price allocations, and accounting requirements.

Determination of Value

The process of determining the value of a broadcasting or cable television property is highly specialized, although certain techniques are generic. The primary methods employed are the income approach, which involves forecasting the revenues and profits attributable to a broadcasting or cable business, and the market approach,[1] which involves analyzing sales of similar businesses and, if appropriate, the behavior of publicly traded media stocks.

---

1. As discussed later in this chapter, the market approach is indicated as the preferred valuation method for assets in certain financial accounting standards such as FAS 142, although this method is not always practical in the case of certain media intangible assets.

In order to determine the fair value of a broadcasting or cable business, projections of discounted cash flow are developed by professional appraisers or by the in-house staff of media companies. The income approach measures the economic benefits the business can reasonably be expected to generate. The fair value of the assets of a cable or broadcasting business may be expressed by discounting these expected future benefits.

Other commonly used valuation methods include the cost approach. For a number of reasons, the cost approach is often not suitable to the valuation of an ongoing broadcasting or cable business because most of their assets are intangible. The principal assets of these businesses are FCC licenses, in the case of broadcasters, and cable franchises, in the case of cable companies. Comparable FCC licenses and franchises cannot be obtained at an identifiable cost. The FCC controls the licensing process, and municipalities have traditionally controlled the cable franchising process; as such, there is no significant marketplace in which similar licenses or franchises can be purchased. Cable franchises were originally granted through a competitive process based on service offerings, and cannot be bought or sold apart from an ongoing business. Consistent with the paradigm of lower of cost or market, the cost approach may be applicable in the case of certain specific assets.

However, the market approach is useful in the valuation of broadcasting and cable businesses. Using marketplace transactions, sales of broadcast stations and cable systems can be expressed as multiples of operating cash flow, which facilitates the comparison of stations or systems of different sizes or in different markets.

Television and radio businesses have typically changed hands at multiples of between 10 and 17 times current operating cash flow, depending upon growth potential, market size, interest rates, and other characteristics. Cable values are expressed as both multiples of cash flow and multiples of subscribers. The average value of cable systems increased from approximately $2,000 per subscriber in the early 1990s to over $5,000 in the early 2000s as investors became attracted to the growth potential of telephony and Internet services. Values have now settled back into the $3,000- to $4,000-per-subscriber range.

A thorough valuation begins with an analysis of the market in which a station or system operates. Figure 16.1 contains a sample page from *The Television Industry: A Market-by-Market Review* (National Association of Broadcasters/Bond & Pecaro, Inc.). This reference book and others like it provide information on important variables such as population and income growth, competition, market share, total market revenues, comparable transactions, and the like. These variables form the initial inputs used to develop the discounted cash flow model, and can also affect assumptions regarding future growth rates and similar factors.

DES MOINES–AMES

DMA: 73

MARKET SUMMARY

	2000	2005	2010	2000–2005 Growth Rate	2005–2010 Growth Rate
Population (000s)	995	1,045	1,076	1.0%	0.6%
Households (000s)	395	414	427	0.9%	0.6%
Retail Sales ($ mill)	N/A	14,317	16,168	N/A	2.5%
EBI ($ mill)	17,455	18,772	21,667	1.5%	2.9%

	Nov. 2000	Nov. 2000	Race (%)		Age (%)	
Cable Penetration	61%	53%	White:	88.9%	<18:	23.4%
VCR Penetration	87%	90%	Black:	2.3%	18–34:	23.5%
ADS Penetration	N/A	25%	Hispanic:	3.4%	18–49:	45.2%
			Other:	5.4%	25–54:	41.4%
					35+	53.0%

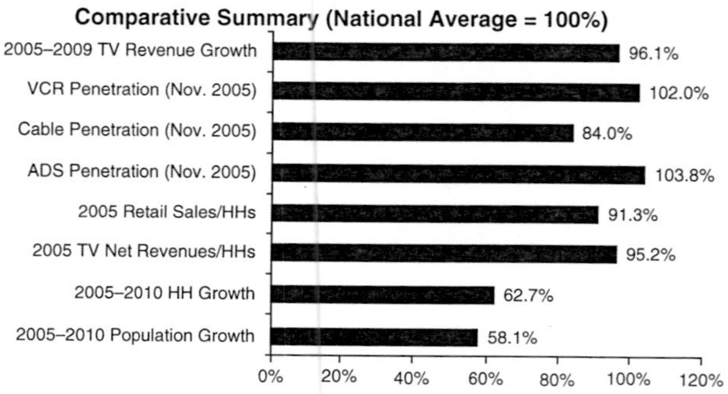

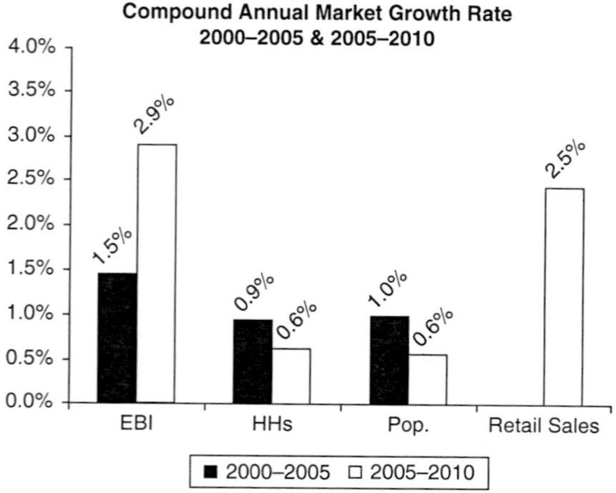

COMPARATIVE SUMMARY	Market Estimate	National Average	% of National Average
2005–2010 Population Growth	0.6%	1.0%	58.1%
2005–2010 HH Growth	0.6%	1.0%	62.7%
2005 TV Net Revenues/HHs	$126	$132	95.2%
2005 Retail Sales/HHs	$34,597	$37,892	91.3%
ADS Penetration (Nov. 2005)	25%	24%	103.8%
Cable Penetration (Nov. 2005)	53%	63%	84.0%
VCR Penetration (Nov. 2005)	90%	88%	102.0%
1999–2004 TV Revenue Growth	N/A	3.4%	N/A
2005–2009 TV Revenue Growth	3.7%	3.9%	96.1%

MAJOR CABLE SYSTEMS AND LOCAL CABLE ADVERTISING

System Name/Owner	Subs. Principal City of Service Area	
Mediacom Communications	90,850 Des Moines	
Mediacom Communications	16,806 Ft Dodge	
Mediacom Communications	8,725 Ames	
2005 Cable Advertising ($000): $8,500	**2005 Cable Adv. as % of Television Advertising:**	**16.3%**

FIGURE 16.1 *Sample Market Profile.*

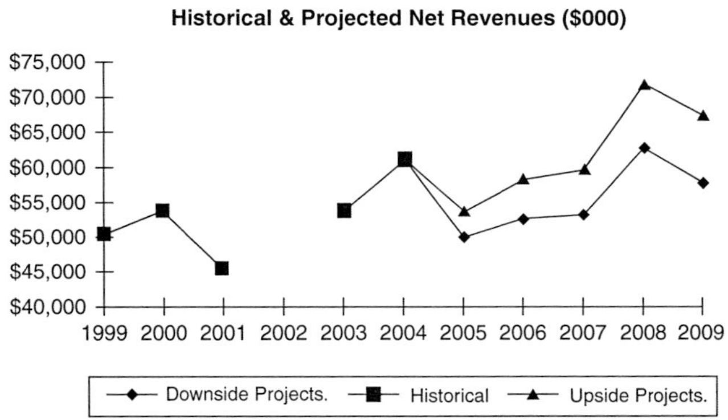

HISTORICAL FINANCIAL DATA

	1999	2000	2001	2002	2003	2004	Average Growth
Network Compen. ($000)	2,380	2,024	1,741	N/A	1,319	1,180	−13.1%
Nat'l/Reg'l Advert. ($000)	16,731	17,663	14,836	N/A	14,653	14,862	−2.3%
Local Advertising ($000)	33,840	34,935	32,932	N/A	38,184	37,791	2.2%
Gross Advertising ($000)	52,936	57,157	48,037	N/A	57,466	65,246	4.3%
Net Revenues ($000)	50,702	53,765	45,481	N/A	54,031	60,875	3.7%
Annual Net Rev. Growth	-------	6.0%	−15.4%	N/A	N/A	12.7%	

MARKET NET REVENUES PROJECTIONS

	2005	2006	2007	2008	2009	Average
Upside Annual Growth Rate	−11.6%	8.4%	2.1%	20.5%	−6.1%	
Upside Net Revenue Estimate ($000)	53,800	58,300	59,500	71,700	67,300	5.8%
Downside Annual Growth Rate	−17.5%	5.2%	0.9%	17.6%	−8.0%	
Downside Net Revenue Est. ($000)	50,200	52,800	53,300	62,700	57,700	3.5%

RECENT TELEVISION SALES

Station	Buyer	Seller	Price ($ mil.)	Date
KDMI-LP	Pappas Telecasting Co.	Caroline K. Powley	1.0	Jan-04

STATION SUMMARY

Calls	Affil.	City of License	Chan.	Start Date	HAAT (ft.)	Visual Power (kW)	# of Sats.	Owner	7 AM–1 AM Audience Share 2002	2003	2004	2005	Adjusted 2005	Prime Time Mon-Sun. Audience Share 2002	2003	2004	2005	Adjusted 2005
WOI	ABC	Ames	5	Feb-50	1850	100	0	Citadel Communications Co. Ltd.	9	9	8	8	13	11	10	11	12	20
KCCI	CBS	Des Moines	8	Jul-55	1939	316	0	Hearst-Argyle Television, Inc.	23	23	24	24	41	21	23	23	23	38
KDIN	PBS	Des Moines	11	Apr-59	1973	316	0	Iowa Public Broadcasting Board	3	3	4	4	N/A	4	4	5	4	N/A
WHO	NBC	Des Moines	13	Apr-54	1970	316	0	New York Times Co.	17	17	18	18	31	17	17	14	15	25
KDSM	FOX	Des Moines	17	Mar-83	2008	3030	0	Sinclair Broadcast Group Inc.	5	5	4	5	8	4	5	5	5	9
KPWB	WBN	Ames	23	Jan-01	2011	5000	0	Pappas Telecasting Companies	4	5	4	3	6	4	4	4	4	7
KFPX	PAX	Newton	39	Nov-98	505	4470	0	ION Media Networks	1	1	0	0	1	2	1	1	1	1
								Total	62	63	62	63		63	64	63	63	
								HUTs	29	30	29	30		55	54	53	53	

DES MOINES–AMES

DMA: 73

FIGURE 16.1 *(Continued)*

The discounted cash flow models used in broadcasting appraisals incorporate variables such as market revenues, revenue-share projections, operating expenses, profit margins, capital expenditures, and various discount rates. In the cable industry, important variables include subscriber growth rates, revenues per subscriber for different services, and churn (percentage of subscribers who disconnect cable service during a specified period).

A discounted cash flow projection period of five to ten years is often determined to be an appropriate time horizon for the analysis. Broadcast stations and cable system investors typically expect to recover their investments within a five- to ten-year period. It is during this time frame that projections regarding revenues, market share, and operating expenses can be made with a degree of accuracy.

The process essentially involves developing a picture of the financial future of a business, with projections of each important variable: market revenues, station or system revenues, operating expenses, income taxes, capital expenditures, and the like. Proper consideration of tax implications is particularly important because, as will be explained below, businesses that rely primarily on large intangible assets enjoy significant tax benefits.

Federal and state income taxes are deducted from projected operating profits to determine after-tax net income. Depreciation and amortization are also added back to the after-tax income stream, and projected capital expenditures and, if appropriate, working capital adjustments are subtracted to calculate a business's net after-tax cash flow.

The stream of annual cash flows is adjusted to present value using a discount rate appropriate for the business. The discount rate used is typically based upon an after-tax rate calculated for the industry. This rate can be derived from information about the stock prices and debt characteristics of industry peers and likely acquiring companies.

Additionally, it is necessary to project the broadcast station's or cable system's terminal value. Two methods are often used in this calculation. An operating cash flow multiple can be applied to a station or system's operating cash flow at the end of the projection period, and then adjustments can be made for capital gains taxes and other expenses to yield the actual proceeds that will accrue to the owner of the station or cable system. Alternatively, an annuity method, also known as the Gordon Growth Model or the perpetuity method, can be used. This entails dividing an estimate of free cash flow, also referred to as net after-tax cash flow, at the end of the forecast period by a capitalization rate. The terminal value represents the hypothetical value of the business at the end of the projection period. The terminal value is then discounted to present value at an appropriate discount rate.

Case Study

Table 16.1 contains an example of a discounted cash flow forecast for a hypothetical television station, WUVW, in a Top 50 market with total market television revenues of just over $100 million.

Several characteristics are noteworthy about the projection in Table 16.1—some of which reflect changes that a likely buyer would implement in the financial performance of the station, and some of which relate to regulatory and tax factors:

1. In this case, the station is seen to have the potential to increase its share of market revenues to 6 percent from 5.2 percent. Typically, this judgment is based upon the performance of similar stations in other markets or the station's actual audience performance. Parameters used in these comparisons include channel position, coverage area, competition, the number and type of television signals (i.e., UHF vs. VHF) in a market, likely network affiliation, and the demographics of the market.
2. There is also an expectation that the operating cash flow can be improved to 27.6 percent from 23.8 percent. The "mature" margin is typically based upon an industry norm derived from analyst reports or from surveys conducted by third-party entities such as the National Association of Broadcasters (NAB) and the Broadcast Cable Financial Management Association (BCFM).
3. It is noteworthy that taxable income is significantly less than operating cash flow for most of the projection period, and that no income taxes are payable until well into the forecast period. This is due to the impact of Section 167 of the Internal Revenue Code. Section 167 permits intangible assets, which account for most of the value of a television or cable business, to be written off against income on a straight-line basis over a 15-year period. (More on this in the discussion of purchase price allocations below.)
4. In each year, the free cash flow is discounted to present value, in this case at a discount rate of 8 percent. For example, the $1,412,000 to be received in Year 3 of the forecast has a present value of only $1,165,000 in today's dollars. In total, the year-to-year after-tax cash flows have a present value of $10,440,600.

As discussed earlier, the investor still owns the broadcast station or cable system at the end of the projection period. In this case, as indicated in Table 16.2, the terminal value was calculated to be $23.8 million. Complex calculations go into the projection of this amount. In this case, the terminal value is equivalent to about 8 times Year 10 operating cash flow and 14 times Year 10 free cash flow. The present value of the terminal value is $11.0 million. When this amount is added to the present value of the cumulative cash flows and a provision for the amortization remaining at the end of the forecast period, a total value of $22,468,000 for the station is indicated.

	YEAR 1	YEAR 2	YEAR 3	YEAR 4	YEAR 5	YEAR 6	YEAR 7	YEAR 8	YEAR 9	YEAR 10
Market Net Revenues	$102,900.0	$105,884.1	$108,954.7	$112,114.4	$115,365.7	$118,711.3	$122,153.9	$125,696.4	$129,341.6	$133,092.5
Sample Television Market Revenue Share	5.2%	5.5%	5.8%	6.0%	6.0%	6.0%	6.0%	6.0%	6.0%	6.0%
Projected Sample Television Revenues	$5,350.8	$5,823.6	$6,319.4	$6,726.9	$6,921.9	$7,122.7	$7,329.2	$7,541.8	$7,760.5	$7,985.6
Operating Profit Margin	23.8%	25.3%	26.8%	27.6%	27.6%	27.6%	27.6%	27.6%	27.6%	27.6%
Operating Cash Flow	$1,273.5	$1,473.4	$1,693.6	$1,856.6	$1,910.4	$1,965.9	$2,022.9	$2,081.5	$2,141.9	$2,204.0
Taxable Income[1]	($568.1)	($735.3)	($233.3)	$97.4	$128.3	$314.0	$510.2	$565.6	$624.4	$684.9
Income Taxes	0.0	0.0	0.0	0.0	0.0	0.0	0.0	31.5	249.8	274.0
After-Tax Income	($568.1)	($735.3)	($233.3)	$97.4	$128.3	$314.0	$510.2	$534.1	$374.6	$410.9
Net After-Tax Cash Flow[2]	$1,020.6	$1,194.2	$1,412.4	$1,582.8	$1,654.2	$1,709.2	$1,765.8	$1,792.4	$1,633.9	$1,671.3
Present Value Net After-Tax Cash Floe @ 8.0%	$982.1	$1,064.0	$1,165.2	$1,209.0	$1,170.0	$1,119.4	$1,070.7	$1,006.3	$849.4	$804.5
Cummulative Present Value Net After-Tax Cash Flow	$982.1	$2,046.1	$3,211.3	$4,420.3	$5,590.3	$6,709.7	$7,780.4	$8,786.7	$9,636.1	$10,440.6

TABLE 16.1 *Projected Sample Television WUVW Operating Performance (Dollar Amounts Shown in Thousands)*

[1]Taxable income is computed as operating cash flow less depreciation and amortization.
[2]Net after-tax cash flow is calculated as after-tax income plus depreciation and amortization less capital expenditures and changes in working capital.

Future Terminal Value in Year 10	$23,811.2
Discounted Terminal Value @ 8.0%[2]	$11,029.2
Total Present Value Cash Flow[1]	10,440.6
Plus: Present Value Tax Benefit of Remaining Amortization	997.7
Fair Market Value Sample Television Station WUVW	$22,467.5

Table 16.2 *Valuation of Sample Television WUVW (Income Method) (Dollar Amounts Shown in Thousands)*

[1] The terminal value is typically calculated through a multiple of operating cash flow or free cash flow, with appropriate adjustments for depreciation and amortization, capital expenditures, working capital, and income taxes.
[2] See text.

Bringing the whole analysis back to today's terms, the value of approximately $22.5 million is equivalent to a multiple of approximately 17 times the $1.7 million of operating cash flow in the first year of the forecast. This amount might seem high relative to the average range of 10 to 14 times, but is consistent with the "upside" growth potential resulting from increasing the revenue share and expanding the operating margin.

The employment of cash flow multiples is a common application of the market approach in the broadcasting and cable industries. Multiples may be less useful, however, in "turnaround" situations when a station or system is underperforming, or if it is overperforming. In other words, a discounted cash flow may yield a value outside the range indicated by "average" income multiples. The analyst should scrutinize the assumptions that underlie the discounted cash flow analysis to determine whether the variance is justified, or whether inputs to the model should be reconsidered. In the case of our example, the indicated value fell above the range, but this appeared to be justified by opportunities to dramatically increase cash flows.

The $22.5 million is the value of the operating assets of the business. This is the most typical configuration for a media acquisition because it is the cleanest. Often, however, the purchaser will acquire the stock of the entity that owns those assets, rather than the assets of the business directly. For example, if BCFM Broadcasting, Inc. acquires Television Station WUVW in an asset purchase from UVW Holdings LLC, it is purchasing only WUVW's equipment, studios, FCC licenses, contracts, and related assets. UVW Holdings would still be owned by its stockholders, just without the operating tangible and intangible assets of WUVW.

In a stock transaction, BCFM Broadcasting, Inc. acquires UVW Holdings as a company, and gets more than just the WUVW operating assets. It typically assumes UVW Holding's debt and other "nonoperating" assets and liabilities, such as cash, accounts receivable and payable, deferred taxes, and the like. Also assumed are any potential lawsuits or claims (tax, environmental, employment, or other claims, both known and unknown) that may be brought against UVW Holdings. The risk related to such contingent liabilities is one of the reasons buyers typically prefer an asset transaction or generally pay a reduced purchase price for a stock transaction.

To arrive at a stock value, the value of the operating assets needs to be adjusted for those other assets and liabilities. Using the example above, the value of UVW Holding's equity is as follows, assuming its only operating asset is Television Station WUVW, and the only other relevant balance sheet items are $6,000,000 in debt and $1,500,000 in cash.

Station Value	$22,500,000
Plus: Cash	1,500,000
Less: Assumed Debt	(6,000,000)
Equals: WUVW 100 percent Equity Value	$18,000,000

Interestingly, however, although asset purchases are typically preferred by the buyer, the formulation of a purchase agreement in an asset transaction may be more burdensome than in a stock deal. This is due largely to the need to carefully itemize the tangible and intangible items that convey, and those that do not.

Due Diligence Considerations

The process of due diligence usually takes place between the signing of a letter of intent for a transaction, the formulation of a purchase agreement, and the actual closing of the transaction. Many of these factors are generic, but others, particularly those that relate to FCC licenses and cable franchises, are unique to these industries. The process of due diligence is typically conducted in conjunction with qualified legal, accounting, and appraisal professionals, and is intended to avoid potentially damaging discoveries after the closing has taken place.

The primary components of due diligences include confirming the accuracy of information and compliance with laws and regulations in areas including:

1. **Organizational Standing**—Particularly if stock is being acquired, making sure that the company is properly and legally organized and incorporated, that the necessary documents are in order, and that necessary fees and taxes are up-to-date.

2. **Licenses and Franchises**—FCC licenses and cable franchises are subject to unique requirements, and it is critical that they be in conformity with FCC and municipal requirements.

3. **Financial Statement Review**—Because the acquirer of a cable or broadcasting business is acquiring a stream of income (hence the prevalence of cash flow multiples in valuation), it is critical that financial statements be analyzed and verified.

4. **Real Property**—Owned realty needs to be properly titled, and copies of all related titles, deeds, mortgages, and the like should be provided. Documentation also needs to be provided—and, in some cases, testing conducted—to ensure that the properties are in compliance with all environmental regulations.

5. **Tangible Assets**—A detailed inventory of acquired assets should be provided and confirmed. In some cases, the acquirer will conduct a detailed fixed asset inventory.

6. **Intellectual Property and Other Intangible Assets**—Particularly important intangible assets at a broadcasting business include programming agreements, talent and management contracts, facilities leases, income leases, and advertising-related assets. Key cable system intangibles may include supplier agreements, programming agreements, multiple dwelling unit (MDU) contracts, and the acquired base of subscribers. These also need to be validated and documented.

7. **Stock Transaction Due Diligence**—The due diligence requirements of a stock transaction are generally more important and complex than in an asset transaction because the buyer is in essence stepping into the seller's corporate shoes and assuming responsibility for more known and possibly unexpected claims related to agreements with prior employees, customers, vendors, benefit plans, and the like.

Due diligence procedures also include identification of going-forward operational strategies such as which contracts will be assumed and which will need to be renegotiated. Other considerations include a staffing assessment, evaluation of talent and news strategies, market research, and consolidation opportunities.

Purchase Price Allocations

Once the total value of a transaction is determined and commitment to the transaction is made by a purchase agreement, it is necessary to embark upon a second type of valuation analysis: the purchase price allocation.

For financial reporting and tax purposes, it is necessary to allocate the total value to different classes of tangible and intangible assets. The treatment of these assets can be quite different for tax and financial reporting purposes. For example, FCC licenses and cable franchises are treated as Section 197 intangible assets, almost all of which are written off over a 15-year life for tax purposes. For accounting purposes, however, they are treated as "indefinite-lived" intangible assets, which simply remain on the balance sheet at their acquisition cost unless a triggering event occurs that causes the value to decline below the acquisition cost, at which time a revaluation of the assets in conformity with FAS 142 (accounting for goodwill and other intangible assets) is conducted.

RADIO		CABLE	
CATEGORY	**VALUE**	**CATEGORY**	**VALUE**
Tangible Assets		Tangible Assets	
Land	$124,539	Land	$164,300
Land Improvements	23,156	Land Improvements	35,444
Buildings	100,000	Buildings	1,000,000
Leasehold Improvements	15,922	Leasehold Improvements	178,433
Towers	934,000	Headend Equipment	1,533,000
Antenna System	76,439	RF Distribution Plant	27,933,000
Transmitter Equipment	97,355	Subscriber Drop Plant	1,944,335
Studio Technical Equipment	495,000	Converters	2,899,784
Microwave Equipment	23,125	Vehicles	115,433
Furniture and Fixtures	198,563	Furniture and Fixtures	1,243,550
Other Tangible Assets	204,566	Other Tangible Assets	4,255,344
Total Tangible Assets	$2,292,665	Total Tangible Assets	$41,302,623
Intangible Assets		Intangible Assets	
Advertiser List	6,553	Favorable Facilities Leases	425,841
Advertiser Relationships	102,455	Income Leases	149,644
Morning Team Talent Contracts	657,000	Acquired Subscriber Base	21,655,433
Tower Income Leases	55,244	Franchise Operating Rights	135,400,000
Favorable Studio Lease	41,233	Miscellaneous Other Intangible Assets and Goodwill	61,066,459
FCC Licenses	8,552,400		
Miscellaneous Other Intangible Assets and Goodwill	292,450		
Total Value	$12,000,000	Total Value	$260,000,000

TABLE 16.3 *Examples of Purchase Price Allocation*

Typically, the tangible assets in a broadcasting or cable acquisition account for a relatively small portion of the purchase price. The more important assets are the FCC licenses, in the case of a television or radio station, and the franchise agreements, in the case of a cable system.

An inventory of the tangible property is typically conducted in order to ensure that items are properly categorized and documented. Between 20 and 30 tangible asset categories may be involved in a broadcast or cable acquisition, such as land, land improvements, leasehold improvements, buildings, technical equipment, antenna equipment, headend equipment, converters, transmitter equipment, microwave equipment, vehicles, and the like. These assets are usually appraised on the basis of the depreciated replacement cost. Items for which an active used-equipment market exists—such as furniture, office machines, and tools—can also be appraised using a market-sales or comparable-sales approach.

Intangible assets, depending upon the asset in question, can be valued using the income, cost, market, or residual approaches.

A simple purchase price allocation example for a radio station and a cable television system appear in Table 16.3.

Accounting Requirements

Several standards promulgated in recent years by the Financial Accounting Standards Board (FASB) place requirements on when and how asset values are recorded. It should be borne in mind that these standards are different from the prevailing tax regulations. These current standards are summarized in the sections that follow.

Statement of Financial Accounting Standards No. 141: Business Combinations

Under FAS 141, all business combinations are to be accounted for using the purchase method of accounting. Previously, the pooling-of-interests method, which essentially consisted of adding together the balance sheet of the merging entities, was also acceptable.

Where it exists, goodwill, defined as the excess of the cost of an acquired entity over the net of the amounts assigned to acquired assets and assumed liabilities, is to be recognized. Other acquired intangible assets are to be separately recognized if: (1) they arise from contractual or other legal rights, regardless of whether those rights are transferable or separable from the acquired enterprise; or (2) in cases where they do not arise from such contractual or legal rights, only if they are separable—that is, capable of being separated or divided from the acquired enterprise.

The pronouncement contains a long list of intangible assets that need to be considered, which adds a layer of complexity to the acquisition accounting process. For example, it does not appear to be permissible to appraise a single asset that covers the business's base of customers. This should be broken out between the customer list, customer contracts, and customer relationships.

A lengthy list of illustrative intangibles that are deemed to be distinguishable from goodwill has been compiled by the FASB. Among the listed intangible assets used by broadcasting and cable businesses are trademarks; trade names; noncompetition agreements; customer lists; customer and supplier relationships; video, audio, and music materials; licensing and royalty agreements; advertising contracts; lease agreements; franchise agreements; operating and broadcast rights; employment contracts; and Internet domain names. Statement 141 specifically provides that the value of an assembled workforce of at-will employees acquired in a business combination is to be included with goodwill.

When reporting a business combination, notes to the financial statements must disclose specific information where it is significant. For intangible assets subject to amortization: (1) the total amount assigned and the amounts assigned to major intangible assets, (2) the residual value, and (3) the weighted average amortization period by major intangible asset class.

Similarly, for intangible assets without a determinable life: (1) the total amount assigned, and (2) the amount assigned to any major intangible asset class.

In the case of goodwill: (1) the total amount of acquired goodwill, (2) the amount that is expected to be deductible for tax purposes, and (3) the amount of goodwill by reporting segment.

The definition of a reporting segment was the subject of considerable debate when FAS 141 was implemented. For example, a radio group might own 100 individual AM and FM stations licensed to 30 markets, grouped into three regions. This raised the question as to whether a "reporting unit" should be each station, a market cluster, or an entire region. Consistent with the way their stations are organized and report their results, many broadcasters group the properties by market cluster, as opposed to individual call letters, on one hand, or by broad regions, on the other. Cable businesses tend to be divided into rational regional definitions.

Statement of Financial Accounting Standards No. 142: Goodwill and Other Intangible Assets

FAS 142 deals with how intangible assets are treated after the acquisition has occurred. Intangible assets with determinable lives are written off over their respective lives.

FAS 142 requires that intangible assets without determinable lives, which include FCC licenses and cable franchises, be reviewed for impairment. This "testing" needs to occur annually, or more frequently if there is a triggering event such as a bankruptcy or pending divestiture.

A recognized intangible asset should be amortized over its anticipated useful life. Such amortization must reflect any anticipated residual value associated with the asset. By contrast, an intangible asset with an indefinite useful life should not be amortized until its life can be determined. Both types of intangible assets must be tested for impairment annually, or more frequently in cases where impairment may have occurred.

Regarding these assets, the appropriate tests differ. For assets with determinable lives, the sum of the undiscounted cash flows over the remaining life of the asset is compared to the carrying value, in accordance with Statement 144; this is often referred to as a recoverability test (see below). If the sum of the undiscounted cash flows does not exceed the carrying value, impairment exists and the fair value must be determined, with a loss equaling the difference between the fair value and the carrying amount recognized. For intangible assets without determinable lives (those that are not subject to amortization), the test consists of comparing the fair value of the asset to its carrying value. Where necessary, an impairment loss equal to the difference must be recognized.

Finally, Statement 142 states that if goodwill and another asset of a reporting unit are tested for impairment at the same time, the other asset shall be tested for impairment before goodwill. If the other asset is impaired, the impairment loss would be recognized prior to goodwill's being tested for impairment.

A noteworthy aspect of FAS 142 is that a company must book impairment if an asset or a reporting unit declines in value. However, there is no corresponding mechanism to book an increase in value over the cost basis if an asset appreciates.

Statement of Financial Accounting Standards No. 144: Accounting for the Impairment of Long-Lived Assets

Long-lived assets typically include assets held for use such as land, buildings, equipment, natural resources, and various intangible assets. Long-lived assets in the broadcasting and cable industries may include customer relationships, leases, long-term facilities contracts, and the like. Assets such as FCC licenses and cable franchises are assets with indefinite lives.

Typically, a long-lived asset is depreciated over time, and is tested for impairment only if there is a triggering event, such as a dramatic decline in the market overall, poor performance of the asset, or the expectation that the asset will be sold.

Interestingly, the impairment test compares an asset's undiscounted future cash flows with the carrying value. If the asset fails the test, then the measurement of the impairment reverts to the difference between the fair value and the carrying value.

Case Study

So, for example, a five-year contract with a key news personality may have been acquired with a television station and given a carrying value of $250,000 with expected amortization over a five-year period. At the end of Year 2, the carrying value is $150,000, assuming a simple straight-line amortization schedule. Assume further that the particular news broadcast shows a steep drop in ratings accompanied by a decline in advertising revenues. If a revised financial analysis shows that the incremental *undiscounted* cash flow attributable to the news personality is greater than $150,000, no impairment would be indicated. However, if the sum of the undiscounted cash flows is less than $150,000, impairment is indicated. In order to measure the impairment, however, the carrying value must be compared to the asset's fair value, which might be measured by a discounted cash flow analysis or perhaps an analysis of comparable talent contracts. If this exercise indicated that the fair value was $50,000, then an impairment charge would result in a $100,000 reduction in income, and an accompanying reduction of $100,000 in the carrying value of the talent contract.

Statement of Financial Accounting Standards No. 157: Fair Value Measurements

FAS 157 endeavors to provide a new and more consistent definition of fair value, which is as follows:

Fair value is the price that would be received to sell an asset or paid to transfer a liability in an orderly transaction between market participants at the measurement date.

This definition contains a slight but significant variation in the traditional definition for fair market value, as well as the definition of "fair value" that was originally articulated in FAS 142. This distinction has to do with the phrase "market participant." The phrase implies that the acquirer would be a typical buyer, not a particular buyer. In most cases, this would be a diversified media, broadcasting, or cable company. What is important is that fair value in this case may not be the absolute highest value that might be paid, for example, by the one company for which an asset has the highest value. In some cases, it may be difficult to justify a purchase price made for strategic reasons because of the lack of similar transactions to support fair value. Then, because annual evaluations are required based on the operating results, some intangible assets could be subject to an impairment charge shortly after the transaction because of the limited operational benefits realized by the buyer.

In Conclusion

This chapter has provided a brief overview of the financial and accounting requirements encountered when making an acquisition in the broadcasting and cable industries. The processes of determining the value of a property, due diligence, purchase price allocations, and accounting compliance are distinctly important, but at the same time interrelated because the establishment of proper values for acquired assets affects all of these areas.

Looking forward, two trends are apparent. First, there has been recognition in the governmental, accounting, and financial communities that the heavy burdens resulting from Sarbanes-Oxley and related accounting standards have in some cases been excessive, and that some relief may be implemented, especially for smaller companies.

On the other hand, there appears to be no deceleration in the profound and complicated changes that are affecting broadcasters and cable operators—including new technologies, increased competition, the Internet, changes in the relationship between television networks and their affiliates, and changes in the advertising economics, to name a few. These changes will ensure that the process of valuing businesses in these sectors will be dynamic in the years ahead, and that related requirements of due diligence, purchase price allocation, and accounting continue to evolve as well.

Finally, there appears to be a movement to further promote the use of the fair value concept in valuation of other assets and liabilities on the balance sheet. This could further complicate the ongoing valuations of assets, and has the potential for requiring technical expertise to assist in the ongoing asset evaluation process because most internal accounting staffs do not have the specialized knowledge to perform these functions.

The Authors

Trila Bumstead

Trila joined NNB (New Northwest Broadcasters, LLC) in February of 1999 and is currently Executive Vice President and Chief Financial Officer for the company. She was first exposed to radio as an audit manager at Deloitte & Touche (Seattle, WA) and worked on the initial public offering for Entercom. After five successful years at Deloitte & Touche with extensive exposure to publicly and privately held companies she was well prepared for the CFO duties at NNB. During her time at NNB she has focused on developing the centralized financial reporting infrastructure, internal controls, capital projects, new station acquisitions/divestitures, revenue generation programs, expense reduction, and overall cluster management. She holds a degree in Accounting from the University of Washington and was a Class of 2003 participant in the National Association of Broadcasters Educational Foundation (NABEF) Leadership Program in Washington, DC.

Samuel D. Bush

Sam has held the positions of Senior Vice President, Treasurer and Chief Financial Officer of Saga Communications, Inc. since September 1997. Saga is a public company (NYSE-SGA) that owns and/or operates broadcast properties in 26 markets, including 59 FM and 30 AM radio stations, 5 TV stations, 4 low power television stations, 3 state radio networks, and 2 farm radio networks. Previously, he served as Senior Vice President in charge of AT&T Capital's Media Lending Group where he was responsible for making in excess of $750 million in loans to the broadcast industry. Sam graduated from Krannert Graduate School of Management at Purdue University with a MS in Management and Finance.

Geoffrey J. Christian

Geoff is a Senior Tax Specialist who joined Dow Lohnes Price Tax Consulting Group LLC in 2001 and became a Member in 2005 and then became Managing Partner of the group in 2006. His practice primarily consists of successful resolution of state and local tax examinations throughout the country for the information and communications industry. These examinations encompass a broad range of issues including unitary/nonunitary, business/nonbusiness, apportionment, intercompany expenses and add backs, filing positions, and constitutional matters. Geoff is also involved with Internal Revenue Service controversies, transaction taxes, sales and use taxes, property taxes, and other tax planning matters.

Mary M. Collins

Mary is President and CEO of the Broadcast Cable Financial Management Association (BCFM) and its Broadcast Cable Credit Association (BCCA) subsidiary. An industry veteran with more than 20 years' experience, Collins joined BCFM/BCCA in January of 2003. Under her leadership, Mary and her staff have increased the Associations' focus on members' needs resulting in a 35 percent increase in BCFM memberships, an average 77 percent growth in registrations at the Associations' Annual Conference, and an 18 percent increase in gross revenues for BCCA. Programs fueling this growth include monthly Distance Learning Seminars featuring highly relevant content from top industry experts; a revived Regional Seminar program which brings educational content closer to members; updates to the Associations' publications—*The Financial Manager*, the *UPDATE*, the *BCCA Credit and Collection Handbook*, *BCFM Industry Guidelines*, and others; new member benefits; an improved BCCA Credit Reporting System featuring a database of industry-specific credit reports for more than 23,000 agencies, advertisers, and media buying services; and a focus on making the Associations a recognized industry leader when working with the advertising community as well as with regulatory agencies such as the FCC, the IRS, and the FCC. Mary earned an MBA in Marketing and Policy Studies from the University of Chicago and received her BA, Magna Cum Laude, from William Smith College in Geneva, NY.

Laura D. Daigle

Laura has been with Clear Channel Communications since July 1992 when it acquired WRBQ AM/FM in Tampa, FL where Laura was Business Manager. After additional acquisitions in the market, she served as Market Controller of a cluster generating in excess of $40 million in revenue annually. In May 1999 she transitioned to the Broadcast Accounting Department as their Operations Specialist and has worked since on acquisition and divestiture accounting, training, auditing, and special projects. Laura graduated from the University of South Carolina with a BA in History and English and earned an MBA from the University of South Florida.

Leslie Hartmann

Leslie began her broadcasting career with Edens Broadcasting in 1986 as Assistant Business Manager. Her career continued with Group W Radio and Bonneville International Corporation in Market Controller positions, and in 1997 she joined Radio One, Inc., where she held the position of Vice President of Finance and Corporate Controller. During her eight years at Radio One, Leslie worked on the company's IPO and was instrumental in the acquisition of nearly 60 radio stations. She joined

Entercom Communications in 2005, where she began her focus on inventory management and business analysis in her current role as Regional Director of Business Analysis. Leslie holds a business degree from the University of California, Santa Barbara, and an MBA from the University of Phoenix. Leslie has been a member of the BCFM since 1986, and completed an Executive board term as Chairperson of BCFM in 2006.

J. Michael Hines, Esquire
Mike is a member of Dow Lohnes and leads the firm's tax practice. His practice at the firm has focused on tax and corporate law, including business transfers and exchanges; Internal Revenue Service and state audit, appeal, and litigation matters; federal and state legislative work and estate tax and family wealth transfer planning. Mr. Hines coordinated television industry tax executive efforts in 2006 to resolve tax consequences of the FCC's changes in frequency and equipment usage for electronic news gathering. He has also worked with industry organizations and coalitions in preparing comments on Internal Revenue Service coordinated issue papers and proposed regulations affecting media industries. He is the author of legislation in Virginia exempting film and video production from Virginia sales and use tax. Mike also represented the taxpayer-publisher in a New York case establishing that a publisher of controlled circulation magazines was eligible for the manufacturing exemption from the New York sales and use tax. He holds degrees from the University of North Carolina at Chapel Hill and Harvard Law (JD).

Laura A. James
Laura A. James is the Senior Vice President of Finance for Lincoln Financial Media Company (formerly Jefferson Pilot Communications Company). The broadcast subsidiary of a Fortune 500 financial service organization, Lincoln Financial Media controls three network-affiliated TV stations, 18 radio stations in five markets, two syndicated radio programs, and a regional sports network which produces and distributes ACCA and SEC collegiate games. Laura joined the company in 1997 as Assistant Vice President/Corporate Controller and was promoted to Vice President/Corporate Controller and IT before being named Vice President of Finance in 2002. Her experience includes leading assessments and implementation of business systems including traffic, general ledger and human resources/payroll; overseeing post-merger integration; responsibility for SOX compliance; as well as working on purchase and divestiture functions for the company. A CPA licensed in the state of Georgia, Laura began her career as a public accountant for KPMG Peat Marwick. She holds a BBA in Accounting from Georgia Southern University.

John E. Kampfe
John is Executive Vice President and Chief Financial Officer of Turner Broadcasting System, Inc. In this capacity, he oversees all domestic and international financial

operations of the company related to its news, entertainment, animation, sports, advertising sales, and distribution units, as well as managing the operations of its properties unit. John serves on the TBS, Inc. executive committee and works closely with TBS, Inc.'s parent company, Time Warner Inc., on financial forecasting and reporting. John began his career at TBS, Inc. in April 1992 as Group Controller for Turner's advertising and cable sales, sports teams and real estate divisions. John earned a BA in Accounting from the Fisher School of Accounting at the University of Florida. He serves on the board of directors of Broadcast Cable Financial Management.

Andrew Kober

Andy joined Bresnan in 1990 as Controller and has been responsible for the accounting and financial reporting functions of the company's U.S. and international operations. In this role, Andy has worked with various domestic and international operating locations to implement policies and procedures, integrate acquired businesses, recommend improvements in systems and procedures, review financial results with local management and ensure timely reporting of financial and tax information to Bresnan's partners, bankers, and the SEC. Prior to joining Bresnan, Andy was a manager with the New York City office of Ernst & Young. During his six years there, he worked with clients in the manufacturing, legal services, media, broadcasting, and cable sectors, including Bresnan. Andy is a CPA and a summa cum laude graduate of Manhattan College, where he received a BS in Accounting and Information Systems.

Bruce Lazarus

Bruce joined Cable Audit Associates (CAA) as Chief Operating Officer in 1998 and was named the company's CEO in 2001. He has more than 20 years of financial management and accounting experience in the media industry, including consulting for the accounting and finance departments of Fox News Channel, launching and serving as Director of Finance for the CNBC television network, and serving as CFO of Request Television. He had previously served as Corporate Controller for Reiss Media and its Request TV operation and held accounting management posts at NBC Sports and RCA. He earned an MBA in Corporate Finance from the City University of New York's Baruch College and a BS in Resources Management from the State University of New York. He is an active member of the Broadcast Cable Financial Management Association.

Calvin Lyles, Jr.

Calvin is Radio Controller at Greater Media, Inc. He oversees the financial operations of the 20 major market radio stations. Calvin has served in this position for six years and previously held the position of Business Manager at the Greater Media radio cluster in Detroit, MI for six years. Within that period, Calvin has experienced the implementation of two separate traffic/billing systems and two separate general ledger systems. Calvin holds a BS in Business Administration from Merrimack College.

Joyce Lueders

Joyce is the Business/Program Manager—WFLA-TV, Tampa, Florida and a 30-year veteran of the local broadcast business. Joyce began her career at WJKS-TV, now WCWJ-TV in Jacksonville, FL where she held positions as Business Manager and Director of Programming. She joined WFLA-TV in 1990 as Business Manager and in 1996 also took on the responsibility of overseeing the local programming efforts of the station and played a major role in starting its syndication efforts. Joyce's involvement with industry organizations includes serving on the BCFM Board of Directors and Executive Committee, becoming Chairperson in 1999. She was instrumental in starting a local chapter of American Women in Radio and Television in Jacksonville and served as a Regional Director. Joyce currently serves on the Programming Committee for BCFM as well as the local board for the Boys & Girls Club. She holds a BBA from the University of North Florida.

William "Rick" Mangum

Rick is the VP Broadcast Accounting at Clear Channel where he oversees the accounting and financial reporting for Clear Channel's radio and TV divisions and supervises a staff of 35 accountants. Responsibilities have included budgeting, forecasting, payroll, and risk management. Prior to taking on those duties, Rick was CFO for KellyUSA, an air force base redevelopment and a component unit of the City of San Antonio. He holds a BBA and an MBA from the University of Texas at Austin and is a CPA.

Timothy S. Pecaro

Tim is a principal and founder of the firm of Bond & Pecaro, Inc., a Washington based consulting firm specializing in valuations, strategic planning, acquisition analysis, asset appraisals, and related financial services for the media, communications, and technology industries. Before the formation of Bond & Pecaro, Inc. in 1986, Tim was a Vice President with Frazier, Gross & Kadlec, Inc. Tim joined in 1980. He was named Manager of the firm's Appraisal Services Division in 1982, he became Director of Appraisal Services in 1983, and Vice President of the firm in 1984. Tim has actively participated in the development, research, and preparation of appraisal reports for owners of radio, television, cable, newspaper, radio common carrier, telecommunications, new media, and Internet properties. He has also developed several research studies and has participated in special research reports for the FCC and the NAB. Tim received a BA in Radio/Television Communication Arts from Monmouth College in 1976.

Fidel R. Quiralte

Fidel was named VP of Finance and Controller of Game Show Network in November 2004. In his current position, he is responsible for all aspects of financial reporting, general accounting, tax, and internal controls. Prior to this assignment he was VP of Finance and Controller at CDG, a Boeing Subsidiary. He has also held several other key positions at ARCO, Los Angeles Times, and AT&T Wireless. He holds a BS in Accounting from California State University, Fresno and an MBA from the University of Southern California.

Ron Rizzuto

Ron is Professor of Finance in the Daniels College of Business at the University of Denver. His BS is in Finance from the University of Colorado and his MBA and Ph.D. are in Finance and Economics from New York University. Ron's specialty teaching and research areas include capital expenditure analysis, mergers and acquisitions, corporate financial planning, and telecommunication finance. Ron is one of the leading authorities on the economics of telecommunication overbuilds in the United States. He has published numerous articles and has conducted numerous seminars in these areas. Ron is also a Senior Fellow at the Magness Institute at the Cable Center. In this role, he conducts training programs, research, and consulting projects focused on the cable telecommunications industry.

John S. Sanders

John is a founding Principal in Bond & Pecaro, Inc., a Washington, DC-based consulting firm founded in 1986. The firm specializes in providing financial, economic, and valuation services, including fair market valuations and purchase price allocation reports, to media and communications companies. John graduated cum laude from Dickinson College with a degree in Economics and International Studies (Honors). He received an MBA from the Darden School of Business Administration at the University of Virginia. He writes articles and speaks frequently at industry meetings regarding media and communications financial issues.

C. Robin Szabo

Robin is President of Szabo Associates, Inc., which was founded in 1971, and is the first and largest collection firm to specialize in media collections. The firm serves the media and entertainment industries both domestically and internationally and has been active in numerous creditor plans, significant media litigation, and bankruptcy cases. He is originally from Jacksonville, FL, and is a graduate of Florida State University where he earned a BS from the College of Business with major areas of study in Accounting and Finance. He joined Szabo Associates in 1975, and served as Vice President from 1982 to 2006. Robin is the publisher and editor of Szabo's "Collective Wisdom," a quarterly educational newsletter of credit and collection articles and tips with a circulation of 9,300.

Anthony A. Vasconcellos

Tony joined Covington, KY-based Regent Communications in September 1998 as Vice President and Chief Financial Officer. From December 2000 until August 2005, he served as Senior Vice President and Chief Financial Officer. In September 2005, he became Executive Vice President and Chief Financial Officer for the company. From October 1991 until joining Regent in 1998, he was employed by LensCrafters, Inc., a highly acquisitive optical retail company, which by 1998 had 800 retail stores and $1.2 billion in revenues. In 1994 he was repatriated and assumed oversight of financial reporting and financial systems for LensCrafters until leaving to join Regent in 1998. From July 1987 to September 1991, Tony served as an auditor for the international accounting firm of Coopers & Lybrand, a predecessor to Price-waterhouseCoopers LLP. Tony is currently Chairman of the Board of Directors of BCFM.

Glossary

This glossary of some of the terms used in this book is somewhat limited by space. Several exhaustive glossaries exist online, notably those by the *Wall Street Journal* and the *New York Times*. Readers who have questions not answered here are invited to slake their financial thirst for monetary terminology at those well-founded founts.

accelerated depreciation. A method of calculating depreciation with proportionately larger dollar amounts allocated during the first year(s) of the worthwhile life of the purchased item. See depreciation.

account. The detailed record of a particular asset, liability, owners' equity, revenue, or expense.

account executive. Another title for the traditional salesperson, someone who sells advertising to various businesses or "accounts."

accounting equation. A mathematical expression reported on a balance sheet used to describe the relationship among assets, liabilities, and owners' equity of the business model. The basic accounting equation states that assets equal liabilities and owners' equity. See balance sheet.

accounts receivable. A financial reporting category representing current assets earned that have been billed to customers but not necessarily collected from them.

accrual. The recognition of revenue when earned or expenses when incurred regardless of when cash is received or disbursed. Accrued revenue and accrued expenses are recorded in the period in which they are earned or incurred.

accrued assets. Assets from revenues earned but not yet collected.

accrued expenses. Expenses incurred during an accounting period for which payment is postponed.

adjustment report. Typically generated out of the accounts receivable system in order to make the necessary general ledger adjustments to revenue; adjustments are necessary when a client is written-off as a bad debt or when it is learned that the original invoice was inaccurate.

agency commission. A discount, historically 15 percent, on broadcast commercial rates provided to advertising agencies representing station clients. The agency bills the client the full amount, while retaining the difference generated by the discount.

aging of accounts. The classification of accounts by the time elapsed (i.e., aging) after the date of billing or the due date. The longer a customer's account remains uncollected or the longer inventory is held, the greater is its realization risk.

amortization. A financial procedure that spreads the cost of an intangible asset over the expected useful life of the asset. Identical in definition to depreciation, except that the purchased items are considered intangible rather than tangible. Amortization is reflected in both the profit and loss account and the balance sheet of a business.

annualize. The procedure of converting any financial activity into a yearly figure.

assets. Items of value that a business owns or is due. Equipment, vehicles, buildings, creditors, money in the bank, and cash are all examples of the assets of a business. Typical breakdown includes *fixed assets, current assets,* and *noncurrent assets.* "Fixed" refers to equipment, buildings, plant, vehicles, and so on. "Current" refers to cash, money in the bank, debtors, and so forth. "Noncurrent" refers to any assets that do not easily fit into the previous categories.

Audience Deficiency Unit (ADU). A unit of measure, typically a Nielsen audience rating point, that is owed an advertiser because a program did not achieve the anticipated (projected) ratings promised to the advertiser. This deficiency is alleviated by offering the client additional commercials (called "make-goods") at no charge.

audit. Typically a review of financial statements or performance activity to determine conformity or compliance with applicable laws, regulations, and/or standards.

bad debt (or bad debt expense). An open account balance or loan receivable that has proved to be uncollectible and is written off. Bad debts can be deducted as expenses against tax liability.

balance sheet. Statement of financial position that shows: Total Assets = Total Liabilities + Owners' Equity. See assets; liabilities; owners' equity.

barter (barter exchange). A transaction in which two businesses exchange items of value with no cash money involved. In broadcasting, "barter" is a programming term referring to a station's acquiring a program from a syndication company by offering commercial avails, rather than cash, to the syndicator. See trade (trade agreement).

best practices. The operational characteristics of successful companies; that is, the recognized best ways to practice business.

billings. Dollar value of advertising sold for a specific time period. When one wants to know how much advertising was sold in for the month of June, one would ask about "June billings."

bottom line. In accounting/finance terminology, net income after taxes. In general, it is an expression as to the end results of something (e.g., the net worth of a corporation on a balance sheet, sales generated from a marketing campaign, or the final decision on most any subject).

breakeven point. The financial threshold at which revenues and costs are equal; a combination of sales and costs that will yield a no-profit/no-loss operation. Knowing the breakeven profit level is useful to decision makers because it provides insight as to the likely economic feasibility of a project.

broadcast cash flow (BCF) margin. This margin is calculated by dividing an individual station's or station group's operating income by the same station's or group's gross or net revenue. This margin focuses totally on the operating performance at a station level, and ignores any corporate overhead, interest, amortization, depreciation, or taxes.

budget. An itemized listing of the amount of all estimated revenue that a given business anticipates receiving, along with a listing of the amount of all estimated costs and expenses that will be incurred in obtaining the above-mentioned income during a given period of time. A budget is typically for one business cycle, such as a year, or for several cycles (for example, a five-year capital budget).

burn rate. The monthly rate at which a company spends its available money (cash reserves). Applying an expected burn rate, a company can predict when it will run out of disposable cash.

capital. Owners' equity in a business; total assets of a business. Sometimes used to mean capital assets (see below), cash or funds.

capital asset. A long-term asset that is not purchased or sold in the normal course of business. Generally, it includes fixed assets such as land, buildings, furniture, equipment, fixtures, and furniture. See assets.

capital budget. The estimated amount planned to be expended for capital items in a given fiscal period. Capital items are fixed assets such as facilities and equipment, the cost of which is normally written off over a number of fiscal periods. The capital budget, however, is limited to the expenditures that will be made within the fiscal period; comparable to the related operating budgets.

capitalization. A statement of capital within the firm—either in the form of money, common stock, long-term debt, or some combination of all three.

cash accounting (cash basis of accounting). A system of accounting in which revenues are recognized when cash is received and expenses are recognized as disbursements are made. No attempt is made to match revenues and expenses to determine income. Most media companies use a more sophisticated accrual accounting method that recognizes expected revenues and expenses. See accrual; expenses; matching principal; revenue.

cash flow. In short, a report that shows the "flow" of money in and out of the business over a period of time. More specifically, cash flow reflects earnings before interest, taxes, depreciation, and amortization. See EBITA; EBITDA.

centralized/decentralized Finance Departments. The type of departmental structure attributed to a Finance Department, wherein a centralized unit has most of its staff working at one master location, and a decentralized unit has staff positioned in various geographic locations throughout the company.

churn. A measure of customer or subscriber attrition. In cable parlance, churn is calculated as the number of subscribers whose service is terminated voluntarily or involuntarily in a month divided by the average subscribers in that month. For example, a system with 100,000 subscribers that lost 2,500 subscribers in a given month would have a churn rate of 2.5 percent. Systems with high churn rates must attract large numbers of subscribers just to maintain their subscriber level. The system in the example would have to add 2,500 subscribers just to maintain the 100,000-subscriber level.

consistency. An accounting principle asserting that the same accounting policies and procedures have been followed from period to period by an organization in the preparation and presentation of its financial statements.

conservatism. A GAAP principal which avoids overstating profit and the value of assets, or understating losses and liabilities. See GAAP.

contra account (contra revenue). An account created to offset another account. An example would be an account called Sales Discounts, which reduces the original value of an advertising schedule placed on a station.

controller (or comptroller). The top managerial and financial accountant in an organization. Supervises the Accounting Department and assists management at all levels in interpreting and using managerial-accounting information.

credit policy. The agreed procedures within a company for awarding credit to customers. In broadcasting and cable, this would entail providing credit to advertisers buying commercial time on stations or networks. In cable operations this policy will also cover credit offered to cable subscribers.

cost approach. A valuation methodology based upon the cost to replace an asset with one of similar function or utility.

current assets. Assets that are easily convertible to cash and are expected to be collected or consumed within a 12-month period; they include such items as cash, accounts receivable, and prepaid expenses or inventory.

current-cost accounting. The valuing of assets, stock, raw materials, and so on at current market value as opposed to its historical cost.

current liabilities. These include bank overdrafts, short-term loans (less than a year), and what the business owes its suppliers. They are termed "current" for the same reasons outlined under current assets.

cyclicality. Typically refers to a business whose revenues or profits can fluctuate substantially based upon economic variations or other factors. In the television industry, in particular, stations generally enjoy a boost in growth in election and Olympic years, followed by slower growth or even a decline in the "off" years.

days sales outstanding (DSO). A measure of how long on average it takes a company to collect the money owed to it. The DSO ratio is calculated by dividing a company's total accounts receivable by the average net sales per day. This gives management the average number of days of sales remaining unpaid from advertisers or subscribers.

debt-service coverage. The borrower's annual net operating income before debt service and taxes divided by the annual debt service. A measure of how safe the loan is to the lender.

debt-to-equity ratio. Total liabilities divided by total equities. Sometimes denominator is simply shareholders' equity and sometimes the ratio is calculated by restricting the numerator to noncurrent debt. This ratio provides a measure of the cushion available to creditors should the firm be forced to liquidate.

depreciation. An accounting procedure that spreads the cost of a tangible asset over the expected useful life of the asset. Depreciation is reflected in both the profit and loss account and the balance sheet of a business. See accelerated depreciation; straight-line depreciation.

discounting. A common method for expressing in today's dollars the value of money to be received in the future. The process of discounting is often viewed as the inverse of compounding. For example, the value of $100 invested at 10 percent today would compound to $110 in one year. Similarly, $110 to be received one year from now, discounted at a 10 percent rate, has a present value of only $100. The present value of money decreases with the length of time required to receive it. Typically, an investor will employ a higher discount rate to the cash flows forecast for a risky investment than for a safe one.

dividends. Payments to the shareholders of a limited company.

due diligence. The thorough investigation of a potential acquisition candidate, such as a radio or TV station. Often used to refer to the investigation of a company for an initial public offering.

earned income. Wages, salaries, professional fees, and other amounts received as compensation for services rendered. Revenue acquired from loans or investments is not included.

EBIT. Earnings before interest and taxes (profit before any interest or taxes have been deducted). See cash flow.

EBITA. Often defined as cash flow. Earnings before interest, taxes, and amortization (profit before any interest, taxes, or amortization has been deducted). See cash flow.

EBITDA. Often defined as cash flow. Earnings before interest, taxes, depreciation, and amortization (profit before any interest, taxes, depreciation, or amortization has been deducted). See cash flow.

enterprise reporting, business intelligence. The regular provision of information to decision makers within an organization. These reports can take the form of graphs, text, and tables, and typically are disseminated through an intranet as a set of regularly updated web pages (or "enterprise portal"). Alternatively, they may be emailed directly to users or simply printed out and distributed.

equity. The value of the business to its owners or primary investors. The difference between the business's assets and liabilities. See accounting equation; balance sheet.

expenses. The outflow of assets (or increases in liabilities) used in generating revenues. Expenses that do not change as activity level changes are called *fixed* *expenses,* whereas those that change as activity levels change are called *variable expenses.* (Variable expenses are zero when there is no activity.)

fair market value. A common measure of value typically defined as the amount in cash or cash equivalents between a willing buyer and a willing seller, both being fully informed and neither being under compulsion.

fair value. According to Statement of Financial Accounting Standards No. 157, "the price that would be received to sell an asset or paid to transfer a liability in an orderly transaction between market participants at the measurement date." Prior to the implementation of FAS 157, the prevailing definition of fair value, as stated in FAS 142, was the "amount at which an asset (or liability) could be bought (or incurred) or sold (or settled) in a current transaction between willing parties, that is, other than in a forced or liquidation sale."

financial statements. Information about a company's economic resources and obligations at a point in time, the results of its activities during a particular period, and its sources and uses of cash during that period. Most financial statements are prepared using a set of common ground rules, which have been developed over a period of many years, and are called Generally Accepted Accounting Principles (GAAP).

fiscal year (FY). A 12-month period during which income is earned and received, obligations are incurred, encumbrances are made, appropriations are expended, and for which other fiscal transactions are recorded. A fiscal year typically is not the same as the conventional calendar year.

fixed assets. Long-term tangible assets held for business use and not expected to be converted to cash in the current or upcoming fiscal year. These assets usually consist of major items such as land, buildings, equipment, and vehicles, but can include smaller items such as tools.

flight. The schedule of commercials purchased by an advertiser. Major advertisers will introduce several flights during the course of a year.

free cash flow. The cash a business yields to an investor after all expenses have been paid, including capital expenditures, working capital, income taxes, and other cash outlays. This is loosely analogous to the earnings of a publicly traded stock.

general ledger. The name for the formal ledger in which entries posted from the individual recording journals of transactions are reorganized into specific financial statement accounts. See account; journal(s).

Generally Accepted Accounting Principles (GAAP). The common set of accounting principles, standards. and procedures that companies use to compile their financial statements. GAAP are imposed on companies so that investors have a minimum level of consistency in the financial statements they use when analyzing companies for investment.

goodwill. The excess of the cost of an acquired firm or operating unit over the current fair market value of the separately identifiable net assets of the acquired unit. Informally the term is used to indicate the value of the company's reputation in the marketplace and the number of regular customers. In broadcasting and cable businesses, goodwill can be attributed to the entity's long-term success in attracting audiences and advertisers.

human resource management software (HRMS). A computerized system enabling employers to input and retain much data on their workforce, including supervisors, training history, promotion and position history, performance reviews, skill and experience levels, technical certifications, and property assigned.

impairment. When the fair value of an asset falls below the value reported on a balance sheet.

income. Excess of revenue and gains over expenses and losses for an accounting period. See revenue; expenses.

income approach. A valuation methodology based upon the value of the income that an asset is forecasted to generate for its owner.

intangible assets. Assets of a nonphysical or financial nature, such as a loan, an endowment policy, or an FCC license. See amortization; tangible assets.

invoice. A term describing an original document either issued by a business for the sale of goods on credit (a sales invoice) or received by the business for goods bought (a purchase invoice).

journal(s). The place or places where transactions are first recorded.

journal entries. A term used to describe the transactions recorded in a journal.

ledger. See general ledger.

liabilities. A probable future sacrifice of economic benefits arising out of present obligations. Examples include loans taken out for the business, and money owed by the business to its suppliers. Liabilities are included on the right-hand side of the balance sheet, and normally consist of accounts that have a credit balance. See balance sheet.

liquidity (liquidity ratio). A measurement of a company's capacity to handle its short-term obligations as they mature. The most common is the *current ratio*, which is calculated by dividing current assets by current liabilities. Liquidity is also referred to as a company's ability to quickly convert assets into cash.

market approach. A valuation methodology based upon the values indicated by actual transactions involving comparable assets.

matching principle. A method of recording, or "matching," expenses with the revenues associated with these expenses in the period in which the revenues are recognized. This is important to determining net profit for the period. See net profit; cash accounting.

materiality. The magnitude of an omission or misstatement of accounting data that misleads financial statement readers. Materiality is judged both by relative dollar amount involved and by the nature of the item. If an item is material, it should be disclosed in the body of the financial statements or in footnotes.

net future value/future value. Value at a specified date of a sum invested at a specified interest rate.

net loss. The value of expenses less sales, assuming that the expenses are greater (i.e., if the profit and loss account shows a debit balance). See net profit.

net present value. Discounted or present value of all cash inflows and outflows of a project or investment at a given discount rate.

net profit. The value of sales less expenses, assuming that the sales are greater (i.e., if the profit and loss account shows a credit balance). See net loss.

Network Affiliate Finance (NAF) Department. A department found within most cable program networks that is responsible for maintaining contracts with local cable systems and monitoring per-subscriber monthly fees paid by these systems in exchange for network carriage.

network revenue (network compensation). In broadcasting, a prescribed compensation fee paid to a network affiliate station by its parent network.

net worth. See equity.

nontraditional revenue (NTR). Identification of revenue coming to a broadcast station that has not been obtained through conventional spot commercial advertising. For example, a station may sell booths for a health fair or sponsor a concert in which participating stations receive a percentage of the gate.

operating cash flow. Typically defined as revenues minus expenses not including nonoperating items such as interest, depreciation, amortization, income taxes, and allocated corporate expenses. A related measure, broadcast cash flow

(BCF), is cash available after corporate expenses. An additional measure, free cash flow, is the actual cash that is available as a return to the owners, after capital expenditures, taxes, interest, corporate overhead, and all other expenses.

outsourcing. The management decision to have certain business functions done by outside contractors rather than by in-house staff. Usually outsourcing is considered a cost-saving measuring, implying that outside vendors can accomplish tasks more economically.

owners' equity. Assets minus liabilities; paid-in capital plus retained earnings of a corporation; partners' capital accounts in a partnership; owners' capital account in a sole proprietorship. See assets; liabilities; retained earnings.

payback analysis. An operation which determines the time frame required to recover the investment in the project. It is similar to breakeven analysis because it focuses on recovery of one's investment. See breakeven point.

posting. The copying of entries from the journals to the ledgers. See general ledger; journal(s).

power ratio. A means to determine how efficiently a station converts ratings into revenue. More specifically, this ratio measures the station's share of audience compared to its share of advertising revenue spent in the market.

pricing systems. Protocols for establishing pricing, such as the pricing of commercials by a station. *Static pricing systems* focus on a predetermined price based on historical performance, whereas *dynamic pricing systems* respond to current market conditions.

profit and loss account. An account which combines revenues and expenses to reveal the current profit or loss of a business; that is, whether a business has earned more than it has spent in the current year.

profit margin. The percentage difference between the cost of a product and the price for which it can be sold.

pro forma accounts (pro forma financial statements). A set of hypothetical statements; that is, statements as they would appear if some event occurred. These may also be unaudited statements prepared in a format for a specific audience such as shareholders or the press.

public files. A set of files detailing a broadcast station's promises versus performance in the areas of programming and public service. These files can be examined, by appointment, by any member of the viewing public. They typically become very popular when the FCC license term is expiring and the company owning the license is applying for renewal.

realized income. The return or profit that is actually earned or collected over a given time period. An example would be realized income on a sale of common stock. Unrealized income would be the market appreciation of a given stock that has not been sold.

reconciling. The process of checking one financial account against another for accuracy (e.g., checking a bank statement against your own records).

rep firm. A company that represents local radio and television stations to national advertisers. Essentially, these companies broker commercial time and take a prescribed commission for their efforts.

residual value. At any time the estimated or actual net realizable value (proceeds less costs) of an asset.

retained earnings. The amount of money held in a business after its owners have taken their share of the profits.

return on assets (ROA). Net income derived from assets. Investors often look at this measurement to help determine the companies in which they want to invest.

revenue. The sales and any other taxable income of a business.

Securities and Exchange Commission (SEC). Agency authorized by the United States Congress to regulate the financial reporting practices of most public corporations.

shareholders. The owners of a limited company or corporation.

straight-line depreciation. A method of calculating depreciation in which the identical (i.e., fixed) amount is written off in each period. See depreciation; accelerated depreciation.

supply chain. The flow of materials, information, and finances as they move in a process from supplier to manufacturer to wholesaler to retailer to consumer. Many organizations are looking to supply chain optimization as a means of gaining competitive advantages.

tangible assets. Assets of a physical nature. Examples include buildings, motor vehicles, plant and equipment, and fixtures and fittings. See intangible assets.

time value of money. The concept that money invested will increase in value over time. See net present value; net future value.

trade (trade agreement). A transaction in which two businesses exchange items of value with no cash money involved. In broadcasting, stations often will acquire automobiles, furniture, restaurant meals, hotels, and contest prizes in exchange for commercial time to suppliers.

Traffic Department. A department within a broadcast facility, cable network or cable system that is responsible for placing and monitoring commercials purchased by advertisers in various programs and day parts. The Traffic Department works closely with the Sales Department, which provides the advertising orders, and the Business Office, which handles billing after the commercials have aired.

triple play. Phrase that characterizes the suite of three services offered by cable systems: video, telephony, and Internet. Some analysts have extended this nomenclature to a "quadruple play," which includes both fixed-line and wireless telephony services, in addition to video and high speed Internet.

Index